On the Cauchy Problem

This is Volume 3 in
NOTES AND REPORTS IN MATHEMATICS
IN SCIENCE AND ENGINEERING
Edited by WILLIAM F. AMES, *Georgia Institute of Technology*

A list of books in this series is available from the publisher on request.

On the Cauchy Problem

Sigeru Mizohata
Department of Mathematics
Faculty of Science
Kyoto University

1985

SCIENCE PRESS
Beijing
The People's Republic of China

ACADEMIC PRESS, INC.
Harcourt Brace Jovanovich, Publishers
Orlando San Diego New York
Austin London Montreal Sydney
Tokyo Toronto

Science Press Rapid Manuscript Reproduction

Responsible Editor Lü Hong

Published by Science Press, Beijing, China

Distribution rights throughout the world, excluding the People's Republic of China, granted to Academic Press, Inc., U.S.A.

Printed in Hong Kong

First published 1985
ISBN 0-12-501660-3
Science Press Book No. 5042–76
Library of Congress Card Catalog Number: 85-63843

Preface

Upon the invitation of Prof. Chi, I stayed at the Wuhan University in China during October—November 1983, and gave a series of lectures on evolution equations. The audience was Chinese mathematicians in this field from all over China. At that time Prof. Chi made a wonderful lecture note written in Chinese. This English version is essentially the same as it. However to make the lecture note understandable to every one who attended to the lecture and others, I reorganized the materials, and added to the original version some comments and also appendix at the end of each lecture. It will be my great pleasure if this lecture note could stimulate young mathematicians.

Finally I wish to express my thanks to Mr. Masayoshi Hata (Kyoto University) who took the pain to type my manuscript in this excellent form.

S. Mizohata

August 1984 *Kyoto University*

Contents

Lecture I. Evolution Equations

The linear partial differential equation of evolution type is defined by

$$\left(\frac{\partial}{\partial t}\right)^m u(x,t) + \sum_{j=1}^{m} a_j(x,t;\frac{\partial}{\partial x}) \left(\frac{\partial}{\partial t}\right)^{m-j} u(x,t) = f(x,t), \tag{1}$$

where t represents the time variable, x ($\in \mathbb{R}^n$) represents the space variable. $a_j(x,t;\frac{\partial}{\partial x})$ is a differential operator:

$$a_j(x,t;\frac{\partial}{\partial x}) = \sum_{\nu} a_{j\nu}(x,t)\left(\frac{\partial}{\partial x}\right)^{\nu},$$

where the suffix ν runs through a finite set. The Cauchy problem for (1) (with the data at $t = 0$) is expressed by imposing

$$\left(\frac{\partial}{\partial t}\right)^j u(x,t)\bigg|_{t=0} = u_j(x) \quad (0 \le j \le m-1). \tag{2}$$

There are well-known three types of evolution equation:

1) Wave equation

$$\left(\frac{\partial}{\partial t}\right)^2 u - \Delta u = 0 \tag{3}$$

2) Heat equation

$$\frac{\partial}{\partial t} u - \Delta u = 0 \tag{4}$$

3) Schrödinger equation

$$i\frac{\partial}{\partial t} u - \Delta u = 0\ . \tag{5}$$

The following equation

$$\left(\frac{\partial}{\partial t}\right)^2 u(x,t) + \Delta u(x,t) = 0 \ , \tag{6}$$

which is a well-known elliptic equation, cannot be regarded as an evolution equation at least when it concerns with the Cauchy problem. To explain its reason, we must define the notion of the well-posedness. The Cauchy problem for the equation (1) is said to be <u>C^∞-wellposed</u>, if (supposing $f = 0$) for any initial data

$$\Psi \equiv (u_0(x),\ u_1(x),\ \ldots,\ u_{m-1}(x)) \in C_0^\infty(\mathbb{R}^n),$$

there exists a unique solution $u(x,t) \in C^\infty(\mathbb{R}^n)$ satisfying (1) and (2). In this definition, the uniqueness in the class C^∞ is already a strong requirement. For the heat equation and Schrödinger equation, the uniqueness in the above sense fails to hold, whereas for the wave equation the uniqueness holds. Let us explain these. Take the heat equation. Put

$$\psi(t) = \begin{cases} e^{-t^{-2}}, & t > 0 \\ 0, & t \le 0, \end{cases}$$

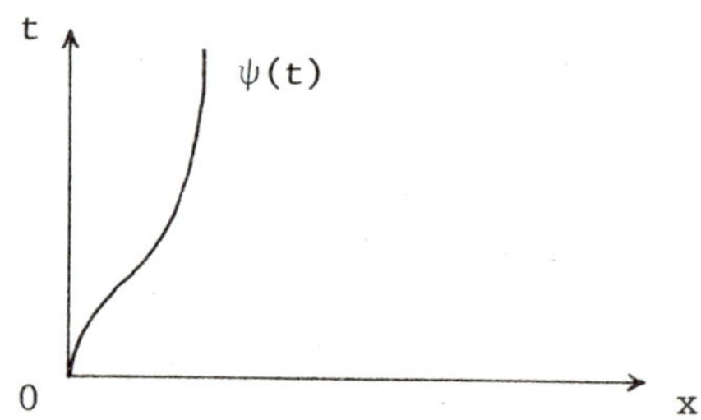

and consider the Cauchy problem

$$\begin{cases} \dfrac{\partial^2 u}{\partial x^2} = \dfrac{\partial u}{\partial t} \\ u(0,t) = \psi(t) = e^{-t^{-2}}, \quad \text{for} \quad t > 0 \\ \dfrac{\partial u}{\partial x}(0,t) = 0 \ . \end{cases}$$

We seek a formal solution

$$u(x,t) = \sum_{j=0}^{\infty} x^j\ \Psi_j(t) \ .$$

We see easily that $\Psi_j(t)$ ($j = 0, 1, 2, \ldots$) are calculated uniquely in terms of $\psi(t)$ and its derivatives, and that this is a convergent series for any x and t, and represents really a C^∞-solution. Now, since $u(x,0) = 0$, $u(x,t) \not\equiv 0$, the solution of Cauchy problem for (4) is not unique. It is the same with (5). Finally, the uniqueness for (3) follows from Holmgren's Theorem.

Now, to the equation (1), we assume

$$a_j(x,t) \in \mathcal{B}^\infty_{x,t} .$$

($\mathcal{B}^\infty$ represents the space of bounded C^∞-functions with all their derivatives.)

The above example shows that in general the uniqueness of the solution cannot be expected unless we impose on the solution $u(x,t)$ some restriction for the its behaviors at infinity. Taking account of this, we define

H^∞-wellposedness. We say that $f(x) \in H^\infty$ if $f(x) \in C^\infty \cap L^2$ with all its derivatives. We say that (1) is H^∞-wellposed for $t \geq 0$, if for all

$$\Psi \equiv (u_0, u_1, \ldots , u_{m-1}) \in H^\infty,$$

there exists a unique solution $u(x,t) \in H^\infty(\mathbb{R}^n_x)$ with $\left(\frac{\partial}{\partial t}\right)^j u(x,t) \in H^\infty(\mathbb{R}^n_x)$ ($0 \leq j \leq m - 1$), $t \geq 0$.

The equations (3),(4),(5) are all H^∞-wellposed for $t \geq 0$. More precisely, (3) and (5) are even H^∞-wellposed for $t \leq 0$, however (4) is not H^∞-wellposed for $t \leq 0$. Observe that (6) is not H^∞-wellposed (ill-posed). This was first observed by Hadamard. Let us remark that, for (6), if the Cauchy data Ψ is analytic, then there exists a solution $u(x,t)$ by the Cauchy-Kowalewski Theorem, and the uniqueness holds by Holmgren. However if the data is in the class C_0^∞, the existence theorem fails to hold. Let us explain this. For (6), we impose the following initial data

$$\begin{cases} u(x,0) = \psi(x) \\ \frac{\partial u}{\partial t}(x,0) = 0. \end{cases}$$

Suppose that there exists a solution $u(x,t)$ for $t \geq 0$. Then extending this solution by $u(x,-t) = u(x,t)$ for $t \geq 0$, we obtain the solution $\tilde{u}(x,t)$ in a full neighborhood of $t = 0$. Now we know that $\tilde{u}(x,t)$ is real-analytic (analytic-hypoellipticity), this implies in particular $\tilde{u}(x,0)$ is analytic in x. This shows that $\psi(x)$ should be analytic.

The essential difference between (3) and (4) is seen by the following property of the corresponding solutions. First for (3), if the initial data has compact support, say contained in $[a,b]$, then the solution $u(x,t)$ has its support in the region shown in the Figure. Sometimes this fact is expressed by saying that " The wave equation has the property of finite propagation speed ". Next, the solution of the heat equation

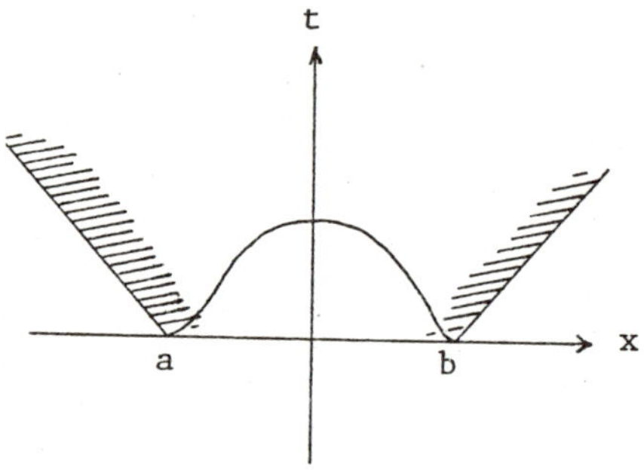

$$\frac{\partial u}{\partial t} = \frac{\partial^2 u}{\partial x^2}, \quad u(x,0) = u_0(x) \in C_0^\infty$$

is expressed by

$$u(x,t) = \frac{1}{2\sqrt{\pi t}} \int_{-\infty}^{\infty} e^{-\frac{(x-y)^2}{4t}} u_0(y)\, dy, \quad t > 0.$$

Now, $u(x,t)$ is analytic in x, and it is easy to see that $\mathrm{supp}[u] = R^2_+$ if $u_0(x) \not\equiv 0$. This shows that the propagation speed of the initial data is $+\infty$. Strictly speaking, this property is not compatible with the reality. However, the heat equation gives a good approximation to the phenomena concerning the heat conduction. Besides this, the heat equation describes the Brownian motion. In this sense also, the heat equation is very important, and the property of its solutions corresponding to some boundary condition is widely investigated. The Schrödinger equation is also important as a fundamental tool in the modern physics.

We finish a brief account of the evolution equations, and let us give a brief sketch of its historical background. Hadamard's well-known treatise [10] gave the decisive orientation and influence to the researches in this field. Before its appearance, there was no definite notion of the well-posedness. At the beginning of the 20th century, only the Cauchy-Kowalewski Theorem was an important theorem in the general theory of partial differential equations. We should also mention the important treatise of Courant-Hilbert [6]. Under the leadership of Courant, in Göttingen, several important research works have been published. In particular, Friedrichs and Lewy were excellent collaborators of Courant. Their methods were based on the so-called L^2-theory. This view-point gained fruitful success.

The view-point of Hadamard and that of Göttingen school were followed by J. Schauder in Poland with success. Let us recall, Schauder was also well-known by the Leray-Schauder Theorem. The third school was in the Soviet Union, and Petrowsky and Sobolev are its representatives. In 1937 and 1938 Petrowsky wrote well-known papers [19],[20] on the evolution equations, especially, on hyperbolic equations (which can be considered as a natural extension of the wave equation). In these works he uses systematically the Fourier transformation. Let us recall that the work of Hadamard relies partly on the notion of the finite part (<u>partie finie</u> in French). Using this notion, he succeeded in constructing the fundamental solution. The work of Petrowsky [19] relies on the Fourier transformation and also difficult techniques, which nowadays could be regarded as those belonging to the pseudo-differential operators. Both works were difficult to understand at that time.

In [19], Petrowsky treated the following equation, which is called today <u>strictly hyperbolic equation</u>. In (1), we suppose

i) $\operatorname{order}_{\xi} a_j(x,t;\xi) \le j \quad (1 \le j \le m).$

ii) Denoting the homogeneous part of degree j of $a_j(x,t;\xi)$ by $h_j(x,t;\xi)$, the characteristic equation

$$\lambda^m + \sum_j h_j(x,t;\xi)\lambda^{m-j} = 0$$

has m real and distinct roots $\lambda_j(x,t;\xi)$ for all non-zero real ξ.

He showed that the strictly hyperbolic equation is C^∞-wellposed.

Now we enter into the 50th. In 1950-51, the well-known treatise by L. Schwartz " Théorie des distributions " appeared. This treatise made clear both the notions of the finite part and of the Fourier transformation and this became a starting point of the researches for young generation. By the way, it seems to the author that Schwartz was noticing the above works of Petrowsky when he wrote the treatise. Schwartz is one of the successors of Hadamard. Of course, we should mention here the famous mathematician J. Leray.

Leray analyzed the work of Petrowsky and also that of L. Gårding just appeared [8], and reorganized in full detail the theory of hyperbolic equations. He gave a series of lectures in Princeton during 1950-54, and his lecture note [15] gave a great influence to the development. In his lecture note, he pointed out a serious mistake in the original proof of Petrowsky. (Now we know that this is not serious if we use the idea of the partition of unity in the dual space [23].)

In 50th, the young generation of Courant school in New York, for instance P. D. Lax and L. Nirenberg, began to publish remarkable results. At that time, Friedrichs and others in Courant Institute were very active [7], [14]. Thus, it was told that there are two big schools in the field of P.D.E., and the study of P.D.E. is monopolized by these two big schools. However we should mention also the Swedish school. L. Gårding has published excellent and elegant works, and L. Hörmander also began to publish new deep results. They are partly influenced by M. Riesz.

Roughly speaking, concerning hyperbolic equations, the works of Leray-Gårding in Europe, and those of Friedrichs and others in New York, were made independently. Namely, the formers usually study higher order equations and the latters study first order systems, especially symmetric hyperbolic systems. We could not see directly the equivalence of those results.

In 1957-58, Zygmund and Calderón in Chicago published papers on singular integral operators [1],[2], and Calderón showed its power by applying their results to the study of the uniqueness of the Cauchy problem.

Entering into the 60th, many mathematicians began to use this flexible tool in the research of delicate problems and recognized its value. Owing to this tool (now is called pseudo-differential operators after remarkable refinements and enlargements), several difficult problems became easier and sometimes even trivial, and several classical, sometimes difficult, works became intelligible for young generation. The pseudo-differential operator enabled young mathematicians to attack delicate (difficult) problems without knowing much the historical background.

Finally we explain more concrete current of the development in the research on hyperbolic equations. As we mentioned above, the case of strictly hyperbolic equations or equivalently the case of strictly hyperbolic systems became familiar among mathematicians. The non-strictly hyperbolic equations could be divided into two categories.

i) the case of variable multiplicities.

ii) the case of constant multiplicities.

Concerning i), in 1958 Chi Min-you published a remarkable result on it [5], and this was analyzed by Oleinik and she discussed the well-posedness for second order equations in fairly general setting [18]. This was again followed by Ivrii-Petkov [12] and Hörmander [11] and others, and now many papers are appearing.

Concerning ii), one knows the fundamental work of E. E. Levi [16], and A. Lax [13] and M. Yamaguti [21] developed the idea of Levi. In 1968, Mizohata-Ohya [17] gave a remarkable result in the case when the multiplicity is at most double. His paper gives a necessary and sufficient condition for the well-posedness, by representing its condition by the vanishing of the sub-principal symbol. From that time, remarkable progress has been made. Flaschka-Strang [24] and Chazarain [3],[4] are well-known. Of course, the case ii) is much easier than the case i).

In this lecture we discuss mainly the Cauchy problem and related topics in the case ii). We discuss also the Cauchy problem in Gevrey class, and also some problems related to the so-called micro-local analysis.

<u>Comments</u>. In this lecture, the knowledge of the basic facts on pseudo-differential operators are assumed. Also, some basic facts about the existence theorem for fairly simple evolution equations are assumed. These facts are found (for instance) in the references in Part II.

References

Part I.

[1] A.P. Calderón and A. Zygmund, Singular integral operators and differential equations, Amer. J. Math., 79(1957), 901-21.

[2] A.P. Calderón, Uniqueness in the Cauchy problem for partial differential equations, Amer. J. Math., 80(1958), 16-36.

[3] J. Chazarain, Opérateurs hyperboliques à caractéristiques de multiplicité constante, Ann. Inst. Fourier, 24(1974), 173-202.

[4] ———, Propagation des singularités pour une classe d'opérateurs à caractéristiques multiples et résolubilité locale, Ann. Inst. Fourier 24(1974), 203-223.

[5] Min-You Chi, On the Cauchy problem for partial differential equations with data given on the degenerate parabolic line, Acta Math. Sinica, 8(1958), 521-29.

[6] R. Courant and D. Hilbert, Methoden der mathematischen Physik, I(1935), II(1937), Springer.

[7] K.O. Friedrichs, Symmetric hyperbolic system of linear differential equations, Comm. Pure Appl. Math., 7(1954), 345-92.

[8] L. Gårding, Linear hyperbolic partial differential equations with constant coefficients, Acta Math., 85(1951), 1-62.

[9] ———, Solution directe du problème de Cauchy pour les équations hyperboliques, Colloques sur les équations aux dérivées partielles du C.N.R.S., (1956), 71-90.

[10] J. Hadamard, Le problème de Cauchy, (1932).

[11] L. Hörmander, The Cauchy problem for differential equations with double characteristics, J. Analyse Math., 32(1977), 118-196.

[12] V.I. Ivrii and V.M. Petkov, Necessary conditions for the correctness of the Cauchy problem for nonstrictly hyperbolic equations, Uspehi Math. Nauk, 29(1974), 3-70.

[13] A. Lax, On the Cauchy problem for partial differential equations with multiple characteristics, Comm. Pure Appl. Math., 9(1956), 135-69.

[14] P.D. Lax, Asymptotic solutions of oscillatory initial value problems, Duke Math. J., 24(1957), 627-46.

[15] J. Leray, Hyperbolic equations (Princeton Lecture Note), 1954.

[16] E.E. Levi, Caratteristiche multiple e problema di Cauchy, Ann. Math. Pure Appl., 16(1909), 109-27.

[17] S. Mizohata et Y. Ohya, Sur la condition de E.E. Levi concernant des équations hyperboliques, Publ. Res. Inst. Math. Sci., 4(1968), 511-26, Sur la condition d'hyperbolicité pour les équations à caractéristiques multiples II, Jap. J. Math., 40(1971), 63-104.

[18] O.A. Oleinik, On the Cauchy problem for weakly hyperbolic equations, Comm. Pure Appl. Math., 23(1970), 569-86.

[19] I.G. Petrowsky, Über das Cauchysche Problem für ein System von partieller Differentialgleichungen, Math. Sbornik, 44(1937), 815-68.

[20] ——————, Über das Cauchysche Problem für ein System linearer partieller Differentialgleichungen im Gebiete der nicht-analytischen Funktionen, Bull. de l'Univ. Etat, Moscow (1938), 1-74.

[21] M. Yamaguti, Le problème de Cauchy et les opérateurs d'intégrale singulière, Mem. Coll. Sci. Kyoto, Ser. A Math., 32(1959), 121-51.

[22] M. Zeman, The Levi-Lax condition for partial differential equations with real characteristics of constant multiplicity, Israel J. of Math., 31(1978), 57-77.

[23] S. Mizohata, Note sur le traitement du problème de Cauchy par des opérateurs d'intégrale singulière, J. Math. Soc. Japan, 11(1959), 213-18.

[24] H. Flaschka-G. Strang, The correctness of the Cauchy problems, Adv. in Math., 6(1971), 347-79.

Part II.

[1] H. Kumano-go, Pseudo-differential operators, (1984), M.I.T. Press.

[2] F. Treves, Introduction to pseudo-differential and Fourier integral operators, 1, 2, (1980), Plenum Press.

[3] S. Mizohata, Theory of partial differential equations, (1973), Cambridge Univ. Press.

Lecture II. H^∞-wellposedness

To understand the basic considerations to the well-posedness, let us explain these in the case when the coefficients a_j of the equation (I,1) depend only on t,

$$\partial_t^m u(x,t) + \sum_{j=1}^{m} a_j(t; \partial_x)\partial_t^{m-j} u(x,t) = f(x,t) \tag{1}$$

where ∂_t, ∂_x stand for $\frac{\partial}{\partial t}$, $\frac{\partial}{\partial x}$ respectively.

The following theorem, due to Petrowsky [1], is well-known. Let

$$\hat{L}_\xi(v) \equiv \partial_t^m v(t;\xi) + \sum_j a_j(t; i\xi)\partial_t^{m-j} v(t;\xi) = 0 . \tag{2}$$

This equation can be regarded as an ordinary differential equation with parameter ξ ($\in \mathbb{R}^n$).

Theorem 1. In order that (1) be H^∞-wellposed with initial data at $t = 0$, and with $f = 0$ in $t \in [0, T]$, the following condition is necessary and sufficient: There exists a constant $C(T)$ and a positive constant p such that for any solution of (2), an estimate of the following form holds:

$$\sum_{j=0}^{m-1} \left|\partial_t^j v(t;\xi)\right| \le C(T)\left(1 + |\xi|\right)^p \sum_{j=0}^{m-1} \left|\partial_t^j v(0;\xi)\right| . \tag{3}$$

Now, in the case f is not zero, namely when

$$t \longrightarrow f(\cdot,t) \in \mathcal{E}_t^0([0, T]; H^\infty)$$

(continuous function in t with values in H^{∞}), we should introduce mor stringent notion of being uniformly H^{∞}-wellposed.

Definition. (uniformly well-posedness)

We say that (I,1) is uniformly H^{∞}-wellposed in the positive direction (or in the future), if for any $\Psi \equiv (u_0, u_1, \ldots, u_{m-1}) \in H^{\infty}$, $t_0 \in [0, T]$, and any $f(\cdot,t) \in \mathcal{E}_t^0([0, T]; H^{\infty})$, there exists a unique solution $u(\cdot,t) \in \mathcal{E}^m([t_0, T]; H^{\infty})$ of (I,1) such that

$$\partial_t^j u \Big|_{t=t_0} = u_j \qquad (0 \le j \le m - 1).$$

Corresponding to this notion, we have the following

Theorem 2. In order that (1) be uniformly H^{∞}-wellposed, it is necessary and sufficient that we have the following estimate (instead of (3)),

$$\sum_{j=0}^{m-1} \left|\partial_t^j v(t; \xi)\right| \le C(T)\left(1 + |\xi|\right)^p \sum_{j=0}^{m-1} \left|\partial_t^j v(t_0; \xi)\right|$$

for $t \in [t_0, T]$, where $C(T)$ and p are independent also of t_0.

From these theorems, we deduce several theorems, which are useful for specific evolution equations. Let us explain the notion of p-evolution equation ($p > 0$). Let us consider (1) in $[0, T]$. Let

$$\max_{1\le j\le m} \ (\operatorname{order}_{\xi} a_j(t; i\xi)/j) = p \quad (\text{rational number}).$$

We assume $p > 0$. The characteristic equation of a p-evolution equation is defined as follows: Let $\mathring{a}_j(t; i\xi)$ be the homogeneous part of degree pj of a_j. Then the equation is

$$\lambda^m + \sum_{j=1}^{m} \mathring{a}_j(t; i\xi)\lambda^{m-j} = 0.$$

Let $\lambda_i(t;\xi)$ $(1 \le i \le m)$ be the characteristic roots of this equation. We have

Theorem 3. In order that (1) be uniformly H^∞-wellposed in the future, the following condition is necessary

$$\mathrm{Re}\,\lambda_i(t;\xi) \le 0 \quad \text{for} \quad {}^\forall(t,\xi) \in [0, T] \times \mathbb{R}^n \quad (1 \le i \le m).$$

Note. This condition is merely a necessary condition. The proof will be given in Appendix.

Now, as an example of the application of Theorem 2, we consider the following equation

$$L(u) \equiv \partial_t^2 u(x,t) - t^{2k}\partial_x^2 u(x,t) - at^s\partial_x u(x,t) = f(x,t) \tag{4}$$

for $t \ge 0$, $x \in \mathbb{R}^1$. $k\ (\ge 1)$ and $s\ (\ge 0)$ are integers, and $a\ (\ne 0)$ is a constant. Note that this equation is not strictly hyperbolic at $t = 0$, although for $t \ge \delta\ (> 0)$ this is strictly hyperbolic. We consider (4) in $t \in [0, T]$. Delicate argument is required in a neighborhood of $t = 0$. In fact, in a neighborhood of $t = 0$, the lower order term $at^s\partial_x u$ cannot be regarded, in general, as a lower order term. To be precise,

Theorem 4. In order that (4) be H^∞-uniformly wellposed for $t \ge 0$, it is necessary and sufficient that

$$s \ge k - 1 \tag{5}$$

Proof.

Sufficiency. Equation (2) becomes

$$\hat{L} v = (\partial_t^2 + t^{2k}\xi^2 + iat^s \xi)v(t;\xi) = 0 . \tag{6}$$

It suffices to show the inequality (3) for ξ positive large, since when we change ξ to $-\xi$, the equation becomes the same if we change a to $-a$. Put

$$\begin{cases} v_0 = \xi v \\ v_1 = \partial_t v. \end{cases}$$

Then, (6) is equivalent to

$$\partial_t V = \begin{pmatrix} 0 & 1 \\ -t^{2k} & 0 \end{pmatrix} \xi V + \begin{pmatrix} 0 & 0 \\ -iat^s & 0 \end{pmatrix} V, \quad V = {}^t(v_0, v_1) \tag{7}$$

$$\equiv H\xi V + BV .$$

The eigen-values of H are $\lambda = it^k$, $-\lambda = -it^k$. Define $N(t) = \begin{pmatrix} -\lambda & -1 \\ -\lambda & 1 \end{pmatrix}$.

Then $N(t)H = \begin{pmatrix} \lambda & 0 \\ 0 & -\lambda \end{pmatrix} N(t) = DN(t) \quad (t > 0)$. Put $N(t)V = W$.

The equation becomes

$$\partial_t W = D\xi W + (N_t'(t) + NB)N^{-1}W.$$

Now $N(t)^{-1} = \frac{-1}{2\lambda}\begin{pmatrix} 1 & 1 \\ \lambda & -\lambda \end{pmatrix}$, and simple computation gives

$$N_t' N^{-1} = \frac{k}{2t}\begin{pmatrix} 1 & 1 \\ 1 & 1 \end{pmatrix}; \quad NBN^{-1} = \frac{iat^s}{2\lambda}\begin{pmatrix} -1 & -1 \\ 1 & 1 \end{pmatrix} = \frac{a}{2t^{k-s}}\begin{pmatrix} -1 & -1 \\ 1 & 1 \end{pmatrix} .$$

Thus

$$\partial_t W = i\begin{pmatrix} t^k & 0 \\ 0 & -t^k \end{pmatrix} \xi W + \frac{k}{2t}\begin{pmatrix} 1 & 1 \\ 1 & 1 \end{pmatrix} W + \frac{a}{2} t^{s-k}\begin{pmatrix} -1 & -1 \\ 1 & 1 \end{pmatrix} W.$$

By assumption, $s - k \geq -1$. Denoting $W = {}^t(w_0, w_1)$, put

$$S(t;\xi) = |w_0(t;\xi)|^2 + |w_1(t;\xi)|^2, \quad t \in (0, 1],$$

$$S'(t;\xi) = 2\mathrm{Re}\ \partial_t w_0 \cdot \bar{w}_0 + 2\mathrm{Re}\ \partial_t w_1 \cdot \bar{w}_1 \leq \frac{\gamma}{t} S(t;\xi),$$

γ being a constant. Hence,

$$S(t;\xi) \leq \exp\left(\gamma \int_{t_\xi}^{t} \frac{ds}{s}\right) S(t_\xi;\xi) = (t/t_\xi)^\gamma S(t_\xi;\xi),$$

which yields by putting $t_\xi = 1/\xi$,

$$S(t;\xi) \leq (\xi t)^\gamma S(1/\xi;\xi), \quad t \in [1/\xi, 1]. \tag{8}$$

The second step is to consider (6), or equivalently (7) for $t \in [0,1/\xi]$. We change t to τ by $\tau = t\xi$. Thus $\tau \in [0, 1]$. Denoting ${}^t(v_0(\tau/\xi;\xi), v_1(\tau/\xi;\xi))$ by ${}^t(\tilde{v}_0(\tau;\xi), \tilde{v}_1(\tau;\xi)) = \tilde{V}(\tau;\xi)$, (7) becomes

$$\partial_t \tilde{V} = \begin{pmatrix} 0 & 1 \\ -t^{2k} & 0 \end{pmatrix} \tilde{V} + \begin{pmatrix} 0 & 0 \\ -iat^s & 0 \end{pmatrix} \frac{1}{\xi} \tilde{V}.$$

We consider this only for $\xi \geq 1$. Since then the coefficients are all bounded for $\tau \in [0, 1]$, we have

$$|\tilde{V}(1;\xi)|^2 \leq \text{const.}|\tilde{V}(0;\xi)|^2.$$ More concretely

$$|v_0(1/\xi;\xi)|^2 + |v_1(1/\xi;\xi)|^2 \leq \text{const.}\left(|v_0(0;\xi)|^2 + |v_1(0;\xi)|^2\right). \tag{9}$$

Finally we return to $S(t;\xi)$. Since $V(t;\xi) = N(t)^{-1}W(t;\xi)$, and that for $t \in [1/\xi; T]$, $|N(t)^{-1}| \le \text{const.}/t^k \le \text{const.}\xi^k$, we have $|V(t;\xi)| \le \text{const.}\xi^k |W(t;\xi)|$. Now by (8),(9),

$$|V(t;\xi)| \le C(T)\xi^{\frac{\gamma}{2}+k} |W(1/\xi;\xi)| \le C'(T)\xi^{\frac{\gamma}{2}+k} |V(1/\xi;\xi)|$$
$$\le C''(T)\xi^{\frac{\gamma}{2}+k} |V(0;\xi)|.$$

This shows that the estimate of the form (3) holds. Now it is easy to show that the inequality in Theorem 2 also holds.

Necessity.

Let us show that when $a \neq 0$, if

$$s < k - 1 \,, \tag{10}$$

the equation (4) is not well-posed. The argument which follows is fairly delicate. The assumption implies that, if we rewrite (6) in the form

$$\hat{L}\, u(t;\xi) = \partial_t^2 u + (t^{2k-s}\xi - ia)t^k \xi u = 0 \tag{11}$$

(writing ia instead of $-ia$), $t^{2k-s}\xi$ becomes weaker (in some sense) than ia. So, we show that there exists a sequence of Cauchy data $v_j(0;\xi)$ and ξ_j (tending to ∞), and t_j (tending to 0) such that the following inequality holds:

$$|v_j(t_j;\xi_j)| + |\partial_t v_j(t_j;\xi_j)| \ge \exp(\delta\xi_j^{\rho})\left(|v_j(0;\xi_j)| + |\partial_t v_j(0;\xi_j)|\right),$$

for some positive constants ρ, δ.

Put

$$\begin{cases} u_0 = t^{\frac{s}{2}} \xi^{\frac{1}{2}} u \\ u_1 = \partial_t u . \end{cases}$$

Then (11) can be written by

$$\partial_t U = \begin{pmatrix} 0 & 1 \\ ia - t^{2k-s}\xi & 0 \end{pmatrix} t^{\frac{s}{2}} \xi^{\frac{1}{2}} U + \frac{s}{2t} \begin{pmatrix} 1 & 0 \\ 0 & 0 \end{pmatrix} U . \tag{12}$$

We define t_ξ (> 0) in such a way that for $t \in [1/\xi, t_\xi]$, we have

$$t^{2k-s}\xi \longrightarrow 0 \quad \text{when} \quad \xi \to \infty.$$

To be precise, we put $t_\xi = \xi^{-\sigma}$ $(0 < \sigma < 1)$. Then $t_\xi^{2k-s}\xi = \xi^{-\sigma(2k-s)+1}$. Hence the above condition is

$$\frac{1}{2k-s} < \sigma < 1 . \tag{13}$$

Under this restriction, we can regard, in (12), the principal part will be $\begin{pmatrix} 0 & 1 \\ ia & 0 \end{pmatrix}$ when ξ is large. The eigen-values of this matrix are $\pm\lambda = \pm\sqrt{ia}$. We can assume $\lambda = \mathrm{Re}\sqrt{ia} > 0$. (In the case when a is of the form $a = ia'$ ($a' > 0$), this is not true. In this case, we replace ξ by $-\xi$ from the beginning.) Define

$$N = \begin{pmatrix} \lambda & 1 \\ -\lambda & 1 \end{pmatrix} \qquad (\lambda = \sqrt{ia}).$$

Then

$$N^{-1} = \frac{1}{2\lambda}\begin{pmatrix} 1 & -1 \\ \lambda & \lambda \end{pmatrix}, \quad N\begin{pmatrix} 1 & 0 \\ 0 & 0 \end{pmatrix}N^{-1} = \frac{1}{2}\begin{pmatrix} 1 & -1 \\ -1 & 1 \end{pmatrix}, \quad N\begin{pmatrix} 0 & 1 \\ ia & 0 \end{pmatrix}N^{-1} = \begin{pmatrix} \lambda & 0 \\ 0 & -\lambda \end{pmatrix}.$$

Thus

$$\partial_t(NU) = \begin{pmatrix} \sqrt{ia} + \varepsilon_{11} & \varepsilon_{12} \\ \varepsilon_{21} & -\sqrt{ia} + \varepsilon_{22} \end{pmatrix} t^{\frac{s}{2}} \xi^{\frac{1}{2}} (NU) + \frac{s}{4t} \begin{pmatrix} 1 & -1 \\ -1 & 1 \end{pmatrix} (NU) \tag{14}$$

where $\varepsilon_{ij}(t;\xi)$ tend uniformly to 0 for $t \in [^1/\xi,\ t_\xi]$ when $\xi \to \infty$.

Put $NU = V = {}^t(\ v_0,\ v_1\ )$, and define

$$S(t;\xi) = |v_0(t;\xi)|^2 - |v_1(t;\xi)|^2 .$$

Then, in view of $\operatorname{Re}\sqrt{ia} > 0$ and $s \geq 0$, simple computation shows

$$\frac{t}{dt} S(t;\xi) \geq \delta t^{\frac{s}{2}} \xi^{\frac{1}{2}} S(t;\xi) \qquad (\ ^{\exists}\delta > 0\ ).$$

By integration,

$$S(t_\xi;\xi) \geq \exp\left(\delta\xi^{\frac{1}{2}} \int_{1/\xi}^{t_\xi} \tau^{\frac{s}{2}}\, d\tau \right) S(^1/\xi;\xi)$$

$$\geq \exp\left(\delta' \xi^{\frac{1}{2}} t_\xi^{\frac{s}{2}+1} \right) S(^1/\xi;\xi)$$

$$= \exp\left(\delta' \xi^{\frac{1}{2} - \sigma(\frac{s}{2}+1)} \right) S(^1/\xi;\xi) \qquad (\ \delta' > 0\ ),$$

if ξ is large. Here we take account of $t_\xi \gg \xi^{-1}$. Now we want to choose σ in such a way that the factor before $S(^1/\xi;\xi)$ becomes infinite when $\xi \to \infty$. For this, it is enough to choose σ in such a way that

$$\sigma(\frac{s}{2} + 1) < \frac{1}{2} \tag{15}$$

Then two conditions (on σ) (13),(15) are compatible, since

$$\frac{1}{2k-s} < \sigma < \frac{1/2}{s/2+1} \tag{16}$$

is nothing but $s < k - 1$, which is our assumption. We fix such a σ. Put $\rho = \frac{1}{2} - \sigma(\frac{s}{2} + 1)$ (> 0). Then the above expression becomes

$$S(t; \xi) \geq \exp(\delta' \xi^{\rho}) S(1/\xi; \xi) \tag{17}$$

if ξ is large.

The last step is to define the initial data (at $t = 0$). Let us recall

$$S(1/\xi; \xi) = |v_0(1/\xi; \xi)|^2 - |v_1(1/\xi; \xi)|^2.$$

We choose the Cauchy data at $t = 1/\xi$ by

$$v_0(1/\xi; \xi) = 1, \quad v_1(1/\xi; \xi) = 0.$$

Hence $S(1/\xi; \xi) = 1$. This implies, since $U = N^{-1}V$,

$$\begin{cases} u_0 = m_{11}v_0 + m_{12}v_1 \\ u_1 = m_{21}v_0 + m_{22}v_1 \end{cases} \qquad (m_{ij} \ : \text{constants})$$

$u_0(1/\xi; \xi) = m_{11}$, $u_1(1/\xi; \xi) = m_{21}$. More concretely

$$\begin{cases} u(1/\xi; \xi) = u_0(1/\xi;\xi)(1/\xi)^{-\frac{s}{2}} \xi^{-\frac{1}{2}} = m_{11}\, \xi^{\frac{s}{2} - \frac{1}{2}} \\ \partial_t u(1/\xi; \xi) = m_{21} \ . \end{cases}$$

This Cauchy data determines the Cauchy data at $t = 0$ ($u(0; \xi)$, $\partial_t u(0; \xi)$) by solving equation (11) backwards. Moreover the same arguments as in the proof of sufficiency show that there exists a constant c_0 (independent of ξ) such that

$$|\xi u(0;\xi)| + |\partial_t u(0;\xi)| \le c_0\left(|\xi u(^1/_\xi;\xi)| + |\partial_t u(^1/_\xi;\xi)|\right)$$

$$\le c' \xi^{\frac{s}{2}+\frac{1}{2}}.$$

This shows that, together with $S(t_\xi;\xi) \ge \exp(\delta'\xi^\rho)$, the Cauchy problem is not well-posed by virtue of Theorem 1. □

Note. From the proof of the sufficiency, we can show easily that in the case $a = 0$, the Cauchy problem for (6) is uniformly well-posed.

Appendix

Proof of Theorem 3.

The following proof is due to Petrowsky. First, without loss of generality, we can assume that at $t = 0$ the condition is violated, namely there exist one of characteristic roots, say λ_1, and ξ^0 ($\in R^n$, $|\xi^0| = 1$), such that

$$\operatorname{Re} \lambda_1(0;\xi^0) > 0 \quad . \tag{18}$$

We want to show that, the forward Cauchy problem for (1) is not well-posed. Put

$$u_j(t;\xi) = |\xi|^{p(m-1-j)}\,\partial_t^{\,j} u(t;\xi) \qquad (0 \le j \le m-1).$$

Putting $U = {}^t(\, u_0, u_1, \ldots, u_{m-1}\,)$, equation (2) is written as

$$\partial_t U = \begin{pmatrix} 0 & 1 & & & \\ & 0 & 1 & & \\ & & 0 & \ddots & \\ & & & \ddots & 1 \\ -\mathring{a}_m(t;i\xi)/|\xi|^{pm} & \cdots & & & \end{pmatrix} |\xi|^p U + \begin{pmatrix} & & \bigcirc & & \\ & & & & \\ * & * & \cdots & \cdots & * \end{pmatrix} U$$

which can be written again as

$$\partial_t U = H_0(t;\xi)|\xi|^p U + BU \tag{19}$$

where $H_0(t;\xi)$ is of homogeneous 0 in ξ, and the entries expressed by $*$ are of order $\leq p'$ ($< p$) in ξ.

Our assumption claims that $H_0(0;\xi^0)$ has one eigen-value $\lambda_1(\xi^0)$ with $\mathrm{Re}\,\lambda_1(\xi^0) > 0$. Let us put $H_0(0;\xi^0) = H$. The following lemma is useful in the treatment of general systems.

<u>Lemma</u> Let H be a matrix of order m with eigen-values $\lambda_1, \lambda_2, \dots, \lambda_m$. Then for any ε (> 0), there exists a matrix N such that

$$NHN^{-1} = \begin{pmatrix} \lambda_1 & & & \\ & \lambda_2 & & \bigcirc \\ & & \ddots & \\ & a_{ij} & & \ddots \\ & & & & \lambda_m \end{pmatrix} \quad \text{where } |a_{ij}| < \varepsilon.$$

<u>Proof of Lemma</u>. If there is no requirement $|a_{ij}| < \varepsilon$, the existence of such an N is well-known. So, we admit this. Denote such one by N_0. Then

$$N_0 H N_0^{-1} = \begin{pmatrix} \lambda_1 & & & \\ & \lambda_2 & & \bigcirc \\ & & \ddots & \\ & a_{ij}^* & & \ddots \\ & & & & \lambda_m \end{pmatrix}.$$

Now denoting

$$I_\varepsilon = \begin{pmatrix} 1 & & & & O \\ & \varepsilon & & & \\ & & \varepsilon^2 & & \\ & & & \ddots & \\ O & & & & \varepsilon^{m-1} \end{pmatrix},$$

We see that

$$I_\varepsilon^{-1} = \begin{pmatrix} 1 & & & & O \\ & \varepsilon^{-1} & & & \\ & & \varepsilon^{-2} & & \\ & & & \ddots & \\ O & & & & \varepsilon^{-(m-1)} \end{pmatrix},$$

Then

$$I_\varepsilon \begin{pmatrix} \lambda_1 & & & O \\ & \lambda_2 & & \\ & & \ddots & \\ & a_{ij}^* & & \ddots \\ & & & \lambda_m \end{pmatrix} I_\varepsilon^{-1} = \begin{pmatrix} \lambda_1 & & & O \\ & \lambda_2 & & \\ & & \ddots & \\ & a_{ij}^{**} & & \ddots \\ & & & \lambda_m \end{pmatrix},$$

where $a_{ij}^{**} = \varepsilon^{i-j} a_{ij}^*$. Since, for $i < j$, $a_{ij}^* = 0$, if we make ε small, $|a_{ij}^{**}|$ become as small as we wish. Thus $N = I_\varepsilon N_0$, for small ε (> 0), satisfies the requirement. □

Now we return to the proof of Theorem 3. Let $\lambda_1, \ldots, \lambda_m$ be the eigen-values of $H_0(0;\xi^0)$ (= H). Without loss of generality, we can assume

$$\operatorname{Re} \lambda_1, \ldots, \operatorname{Re} \lambda_k > 0; \quad \operatorname{Re} \lambda_{k+1}, \ldots, \operatorname{Re} \lambda_m \le 0 . \tag{20}$$

Put

$$\min_{1\le i\le k} \operatorname{Re} \lambda_i = \delta \ (> 0).$$

Now we apply Lemma to H, by taking $\varepsilon = \frac{\delta}{8m}$. We take $\xi = |\xi|\xi^0 = \rho\xi^0$, $\rho > 0$, tends to ∞. Then

$$\partial_t(NU) = D_\varepsilon|\xi|^p(NU) + B'(t;|\xi|\xi^0)(NU) \tag{21}$$

where D_ε is of the form

$$D_\varepsilon = \begin{pmatrix} \lambda_1 & & & \\ & \lambda_2 & & O \\ & & \ddots & \\ a_{ij} & & & \lambda_m \end{pmatrix} + \begin{pmatrix} \tilde{a}_{ij}(t) \end{pmatrix},$$

where $|a_{ij}| \le \frac{\delta}{8m}$, and $\tilde{a}_{ij}(t)$ is continuous with $\tilde{a}_{ij}(0) = 0$. So, if we restrict $t \in [0, t_0]$, $\tilde{a}_{ij}(t)$ $(1 \le i,j \le m)$ become small with t_0. By taking t_0 small, we can assume also $|\tilde{a}_{ij}(t)| \le \frac{\delta}{8m}$. Thus, we have

$$D_\varepsilon = \begin{pmatrix} \lambda_1 & & & \\ & \lambda_2 & & O \\ & & \ddots & \\ O & & & \lambda_m \end{pmatrix} + \begin{pmatrix} a_{ij}(t) \end{pmatrix}, \text{ with } |a_{ij}(t)| \le \frac{\delta}{4m},$$

for $t \in [0, t_0]$. Next, $B'(t;|\xi|\xi^0)$ is estimated, as a function of ρ $(= |\xi|)$, by const.$\rho^{p'}$. Denoting $NU = {}^t(\, v_1(t;\xi), \ldots, v_m(t;\xi)\,)$, define

$$S(t;\rho) = \exp(-\delta\rho^p t)\left[\ \sum_{i=1}^{k}{}' |v_i(t;\rho\xi^0)|^2 - \sum_{i=k+1}^{m}{}'' |v_i(t;\rho\xi^0)|^2\ \right].$$

Let us show that for ρ $(= |\xi|)$ large ,

$$\exp(\delta\rho^p)\frac{d}{dt}S(t;\rho) \ge \frac{\delta}{4}\rho^p \sum_{i=1}^{m} |v_i(t;\rho\xi^0)|^2. \tag{22}$$

In fact,

$$\frac{d}{dt} S = \exp(-\delta\rho^p t)\left(2\mathrm{Re}\ \sum{}' \partial_t v_i \cdot \overline{v}_i - 2\mathrm{Re}\ \sum{}'' \partial_t v_i \cdot \overline{v}_i \right.$$

$$\left. -\delta\rho^p \sum{}' |v_i|^2 + \delta\rho^p \sum{}'' |v_i|^2 \right) .$$

Since

$$\partial_t v_i = \rho^p \left(\lambda_i v_i + \sum_{j=1}^{m} a_{ij}(t) v_j \right) + \sum_{j=1}^{m} b_{ij}(t;\rho\xi^0) v_j ,$$

with $|a_{ij}(t)| \le \frac{\delta}{4m}$, $|b_{ij}(t;\rho\xi^0)| \le \text{const.}\ \rho^{p'}$ ($p' < p$), simple estimates show (22). Thus we have

$$\frac{d}{dt} S(t;\rho) \ge \frac{\delta}{4}\rho^p S(t;\rho).$$

This yields,

$$S(t;|\xi|\xi^0) \ge \exp\left(\frac{\delta}{4}|\xi|^p t\right) S(0;|\xi|\xi^0).$$

Now, taking account of $S(0;|\xi|\xi^0) = \sum{}' |v_i(0;|\xi|\xi^0)|^2 - \sum{}'' |v_i(0;|\xi|\xi^0)|^2$ we see that if we assign $v_1(0;|\xi|\xi^0) = 1$, $v_i(0;|\xi|\xi^0) = 0$ $(2 \le i \le m)$, we have always $S(0;|\xi|\xi^0) = 1$. Thus

$$\sum_{i=1}^{m} |v_i(t;|\xi|\xi^0)|^2 \ge \exp(\delta|\xi|^p t) S(t;|\xi|\xi^0) \ge \exp\left(\frac{5}{4}\delta|\xi|^p t\right).$$

If we return to u_j $(0 \le j \le m-1)$ via N, we see easily that the inequality of the form (3) does not hold.

Comments .

What about the extension of the above method to the case when the coefficients depend also on x ? The next lecture treats this problem. We briefly explain the method. In 1961, the author proposed this [2]. The idea is the following. We localize the operator in x and ξ . To be precise, let (x_0, ξ^0) be a point in $\mathbb{R}_x^\ell \times \mathbb{R}_\xi^\ell \setminus 0$. Let $\beta(x) \in C_0^\infty$ $(0 \le \beta(x) \le 1)$ with small support around x_0, which takes the value 1 in a neighborhood of x_0.

Instead of $u(x,t)$, we consider $\beta(x)u(x,t)$. This consideration has been done from long time ago in the investigation of local structures of solutions.

Next we consider the localization in the ξ-space. For this purpose, we take $\alpha(\xi) \in C_0^\infty$ $(0 \le \alpha(\xi) \le 1)$ with small support around ξ^0, which takes the value 1 in a neighborhood of ξ^0.

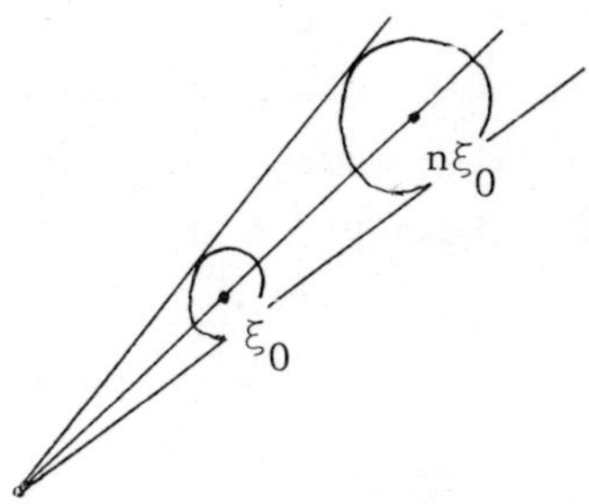

Let

$$\alpha_n(\xi) = \alpha(\xi/_n),$$

where n is a positive integer, or positive parameter tending to ∞.

$\alpha_n(\xi)$ $(n > 0)$ has its support in a fixed conical neighborhood of ξ^0.

Now, we consider $\alpha_n(D)u(x,t)$. This is defined by Fourier transform: $\widehat{\alpha_n u(\cdot,t)} = \alpha_n(\xi)\hat{u}(\xi,t)$. This could be interpreted as "wave packet". We consider, instead of the solution $u(x,t)$, the function $\alpha_n(D)\beta(x)u(x,t)$.

Let $u(x,t)$ be the solution of

$$\partial_t u - a(x,t;\, D_x)\, u = 0.$$

Then operating $\alpha_n(D)\beta(x)$ (micro-localizer) from left, we have

$$\partial_t(\alpha_n(D)\beta(x)u(\cdot,t)) - a(x,t;\, D_x)(\alpha_n(D)\beta(x)u(\cdot,t)) = f_n,$$

where $f_n = [\alpha_n(D)\beta(x),\, a(x,t;\, D_x)]$ ($[\ ,\ ]$ stands for commutator). By this operation, we can say that the operator $a(x,t;\, D_x)$ is localized (if t is small) in x and ξ in a neighborhood of x_0, and a conical neighborhood of ξ^0. If we can show that f_n (arising from the micro-localization) is in some sense negligible, then we can argue in the same way as above by using some energy inequality.

Let us look at f_n.

$$f_n \sim \sum_{|\mu+\nu|\geq 1} (-1)^{|\nu|}\mu!^{-1}\nu!^{-1}a^{(\nu)}_{(\mu)}(x,t;\, D)\left(\alpha_n^{(\mu)}\beta_{(\nu)}\, u\right).$$

This asymptotic expression of f_n gives rise to another micro-localized functions $\alpha_n^{(\mu)}\beta_{(\nu)}\, u$. This requires further considerations. Let us observe that, although this consideration gives us rich results, it has its own limit, because sometimes f_n is no longer negligible as the following example shows. Let us consider the following Schrödinger type equation.

$$\begin{cases} i\partial_t u + \Delta u + \sum_j b_j(x)\partial_{x_j} u + c(x)u = 0 \\ u\Big|_{t=0} = u_0(x) \in H^\infty. \end{cases}$$

If the above method worked, we would have the necessary condition for the H^∞-wellposedness:

$$\mathrm{Re}\, b_j(x) \equiv 0 \quad (1 \leq j \leq n).$$

However this is not true. In fact, in the case $n = 1$, by simple consideration, we get the sufficient condition:

$$\left| \mathrm{Re} \int_0^x b(y)\, dy \right| \quad \text{is bounded.}$$

This condition is no longer of local character. Lecture VII will give a detailed account of this problem.

References

[1] I.G. Petrowsky, Über das Cauchysche Problem für ein System linearer partieller Differentialgleichungen im Gebiete der nicht-analytischen Funktionen, Bull. de l'Univ. de l'Etat de Moscow, (1938), 1-74.

[2] S. Mizohata, Some remarks on the Cauchy problem, J. Math. Kyoto Univ., 1(1961), 109-27.

Lecture III. Lax-Mizohata Theorem

§ 1. Let $P(x,t;\ \partial_x,\partial_t)$ be Kowalewskian. Namely ,

$$P = \partial_t^{\,m} + \sum_{j=1}^{m} a_j(x,t;\ \partial_x)\partial_t^{\,m-j} \tag{1}$$

where order $a_j \le j$ $(1 \le j \le m)$. The coefficients are supposed to be in C^∞ and we consider local Cauchy problem in a neighborhood of the origin. To be precise, we say that the Cauchy problem

$$Pu = f, \quad \partial_t^{\,j} u\Big|_{t=0} = u_j(x) \quad (0 \le j \le m-1) \tag{2}$$

is well-posed in a neighborhood of the origin, or shortly locally uniquely solvable at the origin, if for any Cauchy data $\Psi \equiv (u_0, \ldots, u_{m-1}) \in C^\infty(\mathcal{O}_x)$, ($\mathcal{O}_x$ being a fixed neighborhood of the origin) and $f \in C^\infty(V_{x,t})$, there exists a unique solution $u(x,t) \in C^\infty(V)$, where V is a neighborhood of the origin, which may depend on (Ψ, f).

P.D. Lax [6] observed the following basic fact.

Theorem 1. Suppose (1) to be locally uniquely solvable at the origin, then there exists a common existence domain V_0 of the origin such that for all $\Psi \in C^\infty(\mathcal{O}_x)$, and $f = 0$, there exists a solution $u(x,t) \in C^m(V_0)$ of (2). Moreover, the linear mapping

$$\Psi \in C^\infty(\mathcal{O}_x) \longmapsto u(x,t) \in C^m(V_0)$$

is continuous.

For the proof, see also Mizohata [7]. We use Baire's category theorem, and Banach's closed graph theorem.

Now consider the characteristic equation of (1),

$$\lambda^m + \sum_{j=1}^{m} h_j(x,t;\xi)\lambda^{m-j} = 0 \qquad (3)$$

$h_j(x,t;\xi)$ being the homogeneous part of degree j of a_j. Let $\lambda_i(x,t;\xi)$ $(1 \le i \le m)$ be the characteristic roots of (3). Then

Theorem 2. (Lax-Mizohata)

If the Cauchy problem (2) is locally uniquely solvable at the origin, then for all $\xi \in \mathbb{R}^\ell$, all the characteristic roots at the origin $\lambda_i(0,0;\xi)$ should be real.

Historical Notes. This theorem was first proved by P.D. Lax in 1957 under the assumption λ_i are simple [6]. Incidentally, this paper contains important materials, especially the Fourier integral operator is totally used in the arguments. Next, in 1961, Mizohata proved this without any restriction on λ_i [7], then, so did Ivrii-Petkow [4]. Hörmander [3] gave another proof. See also Kumano-go [5]. Author's proof of 1961 is based on the idea of a kind of micro-localization of operators, which was explained briefly at the end of Lecture II. The method consists of considering micro-localized solution $\alpha_n(D)\beta(x)u$. Our argument is not simple, but the idea is simple. Our proof is fairly lengthy, but the method can be applied to fairly wide range of problems. Indeed, as we shall explain in Lecture V-VI, this method can be applied to the treatment of micro-local analysis in Gevrey class.

Let us first consider the following evolution equation

$$\begin{aligned} &\partial_t u(x,t) - a(x,t;\ D_x)u(x,t) = 0 \\ &u\Big|_{t=0} = u_0(x) \in H^{\infty}, \end{aligned} \tag{4}$$

where $a(x,t;\ D_x)$ is a pseudo-differential operator of order m, m being a positive number (not necessarily positive integer). More precisely, we assume

$$a(x,t;\ \xi) \in S^m_{1,0}\ .$$

This means $a(x,t;\ \xi) \in C^{\infty}(\mathbb{R}^{\ell} \times \mathbb{R}^{\ell})$ and satisfies

$$\left| a^{(\alpha)}_{(\beta)}(x,t;\ \xi) \right| \le C_{\alpha\beta}(1 + |\xi|)^{m-|\alpha|},$$

where $a^{(\alpha)}_{(\beta)} = \partial_{\xi}^{\alpha}(i^{-1}\partial_x)^{\beta}a(x,t;\ \xi) \equiv \partial_{\xi}^{\alpha}D_x^{\beta}a$. t is considered as parameter, and we assume $a(x,t;\ \xi)$ depends smoothly on t for $t \in [0,\ T]$. To discuss the wellposedness, it would be convenient to assume

$$a = a_m + b,$$

where $a_m(x,t;\ \xi)$ is essentially homogeneous of degree m in ξ, and $b \in S^{m'}_{1,0}$ with $m' < m$. The terminology "essentially homogeneous" means that to be homogeneous for $|\xi| \ge A$ (sufficiently large). We call a_m the principal symbol of a.

Let us assume that

$$\mathrm{Re}\ a_m(x,t;\ \xi) \le -\delta|\xi|^m \quad ({}^{\exists}\delta > 0) \tag{5}$$

for sufficiently large $|\xi|$. Then it is not difficult to see that the

Cauchy problem (4) is H^∞-wellposed. Its proof is left to the reader. Our aim is to show that

Theorem 3. In order that (4) be H^∞-wellposed, the condition

$$\operatorname{Re} a_m(x,0;\xi) \le 0, \qquad (x,\xi) \in \mathbb{R}^\ell \times \mathbb{R}^\ell\backslash 0$$

is necessary.

Observe that this theorem is an extension of Theorem 3 in Lecture II. The important fact is to use Banach's closed graph theorem.

Proposition. If (4) is H^∞-wellposed for $t \in [0, T]$, then the linear mapping

$$u_0(x) \in H \longmapsto u(\cdot,t) \in \mathcal{E}_t^0([0, T]; H^\infty)$$

is continuous.

In particular, we have the following fact: There exist a positive constant $C(T)$ and a non-negative integer k such that

$$\sup_{t\in[0,T]} \| u(\cdot,t)\|_{L^2} \le C(T) \| u(\cdot,0)\|_{H^k} . \tag{6}$$

First, observe that in the case where the symbol $a(x,t;\xi)$ depends only on t, Theorem 3 follows immediately from Theorem 1 in Lecture II. In fact,

$$\hat{L}\, v(t;\xi) = (\partial_t - a(t;\xi))v(t;\xi) = 0$$

implies

$$v(t;\xi) = \exp\left(\int_0^t a(s;\xi)\, ds\right) v(0;\xi).$$

Then, if there exists a ξ^0 ($|\xi^0| = 1$) such that $\operatorname{Re} a_m(0;\ \xi^0) > 0$, then by taking $\xi = |\xi|\xi^0$, we have

$$|v(t;|\xi|\xi^0)| \geq \exp(\delta|\xi|^m t)\,|v(0;|\xi|\xi^0)|,$$

if $|\xi|$ is large, and t (> 0) is small, which shows H^∞-illposedness.

Before entering into the proof of Theorem 3, let us give some remarks on our method. We introduce two kinds of cut-off functions $\alpha_n(\xi)$ and $\beta(x)$ as explained in Lecture II. Now,

$$\beta(x)a(x,D)\alpha_n(D)\, u = (2\pi)^{-1}\int e^{ix\xi}\beta(x)a(x,\xi)\alpha_n(\xi)\hat{u}(\xi)\, d\xi \tag{7}$$

$$\left(\ \hat{u}(\xi) = \int e^{-ix\xi}u(x)\, dx\ \right).$$

This means that the symbol $a(x,\xi)$ is transformed into $\beta(x)a(x,\xi)\alpha_n(\xi)$. This implies, in particular, that the operator defined by (7) remains invariant when we change the symbol $a(x,\xi)$ outside the support of $\beta(x)$ (in the x-variable), and also outside the support of $\alpha_n(\xi)$ (in the ξ-variable) In other words, the operator $\beta\, a(x,D)\alpha_n$ is determined only by the symbol $a(x,\xi)$ for $(x,\xi) \in \operatorname{supp}[\beta] \times \operatorname{supp}[\alpha_n]$. We could say that this operator is an micro-localized operator in a neighborhood of x_0, and in a conical neighborhood of ξ^0 (see the definition of $\alpha_n(\xi)$).

§ 2. We explain here some basic facts concerning pseudo-differential operators, which we use here. Appendix will give a brief proof. Let $a(x, \xi) \in S^m_{1,0}$ whose principal symbol, essentially homogeneous of degree m, is denoted by $a_m(x, \xi)$.

1) Estimate from below.

Let

$$\mathrm{Re}\ a_m(x, \xi) \geq \delta|\xi|^m \qquad ({}^{\exists}\delta > 0),$$

if $|\xi|$ is large, in $(x, \xi) \in \mathbb{R}^{\ell} \times V_{\xi^0}$, and suppose that $\mathrm{supp}[\alpha(\xi)]$, therefore $\mathrm{supp}[\alpha_n(\xi)]$ is contained in the conical neighborhood V_{ξ^0} of ξ^0. Then by sharp Gårding inequality, we have, for m > 0, m > 0,

$$\mathrm{Re}\Big(a(x,D)\alpha_n(D)v,\ \alpha_n(D)v \Big) \geq \frac{\delta}{2}\big((1 - \varepsilon')n\big)^m \|\alpha_n v\|^2 \tag{8}$$

if n is large, where we suppose $\mathrm{supp}[\alpha(\xi)] \subset \{\xi; 1-\varepsilon' \leq |\xi| \leq 1+\varepsilon'\}$. The same inequality holds if we replace α_n by $\alpha_n^{(p)}$.

2) Estimate from above.

Let $a(x, \xi) \in S^m_{1,0}$, then we have

$$\| a(x,D)\alpha_n(D)v\| \leq \mathrm{const.}\ n^m \|\alpha_n(D)v\| \tag{9}$$

if n is large. The same inequality holds, when we replace α_n by $\alpha_n^{(p)}$.

3) Estimate of $\|\alpha_n(D)a(x,D)v\|$.

Let $a(x, \xi) \in S^m_{1,0}$, then we have

$$\| \alpha_n(D)a(x,D)v\| \le \text{const. } n^m \| v\| \tag{10}$$

if n is large. If we replace α_n by $\alpha_n^{(p)}$, we have

$$\| \alpha_n^{(p)}(D)a(x,D)v\| \le \text{const.} |\alpha_n^{(p)}(\xi)|_0 \cdot n^m \| v\|$$

$$\le \text{const. } n^{m-|p|} \| v\| . \qquad (10)_p$$

The above inequalities are deduced from (9), by considering the commutator $[\alpha_n^{(p)}, a]$. More precisely, the asymptotic expression of its commutator gives

$$\left\| \left(\alpha_n^{(p)} a(x,D) - \sum_{|\nu| \le N} \nu!^{-1} a_{(\nu)}(x,D)\alpha_n^{(p+\nu)}\right) v \right\| \le \text{const. } n^{m-|p|-N-1} \| v\| . \tag{11}$$

4) <u>Asymptotic expression of $\beta(x)a(x,D)$</u>.

Let

$$\beta(x)a(x,D) = \sum_{|\nu| \le N} (-1)^{|\nu|} \nu!^{-1} a^{(\nu)}(x,D)\beta_{(\nu)}(x) + r_N(x,D).$$

Then $r_N(x,\xi) \in S^{m-N-1}_{1,0}$. Thus, by $(10)_p$, we have

$$\| \alpha_n^{(p)} r_N(x,D)v\| \le \text{const. } n^{m-N-1-|p|} \| v\| . \tag{12}$$

<u>Note</u>. When $a(x,\xi) \in S^m_{1,0}$ and <u>m is negative</u>, one might imagine that $a(x,D)$, as an operator acting on the L^2-space, is a nice operator. However, in general, <u>quantitatively</u> we can not say that its operator-norm $\| a(x,D)\|$ is small. (9) and (10) could be considered as expressing this property. (11) explains this property more precisely.

§ 3. Proof of Theorem 3.

The argument is carried out in the following way. We assume that there exists an (x_0, ξ^0) $(|\xi^0| = 1)$, such that

$$\mathrm{Re}\ a_m(x_0, 0; \xi^0) > 0$$

and that (4) is H^∞-wellposed. From these two assumption, by defining a sequence of initial data suitably, we deduce two inequalities which are not compatible.

Without loss of generality, we assume $x_0 = 0$. Then by taking $\mathrm{supp}[\beta]$, $\mathrm{supp}[\alpha]$ both small, we can assume

$$\mathrm{Re}\ a_m(x,t;\xi) \geq \delta|\xi|^m \quad (\exists\delta > 0) \tag{13}$$

for $(x, \xi) \in \mathrm{supp}[\beta] \times \mathrm{supp}[\alpha]$, and t small, say $t \in [0, t_0]$.

(1st step.) Micro-localization.

Operate $\beta(x)$ to

$$L(u) = \big(\partial_t - a(x,t;D)\big)u(x,t) = 0, \tag{4}$$

(hereafter we write D instead of D_x), then

$$\partial_t(\beta u) - \beta(x)a(x,t;D)\,u = 0.$$

If necessary, by modifying the symbol $a(x,t;\xi)$ outside the support of $\beta(x)$, denoted by $\tilde{a}(x,t;\xi)$, we can assume (13) with $\delta/2$, for $\tilde{a}(x,t;\xi)$, for all $x \in \mathbb{R}^\ell$ and $\xi \in V_{\xi^0}$. Thus, we have

$$\partial_t(\beta u) - \beta(x)\tilde{a}(x,t;D)\,u = 0.$$

Next we operate α_n to obtain

$$\partial_t(\alpha_n \beta u) - \alpha_n \beta \tilde{a}(x,t;D)\, u = 0.$$

By commuting $\alpha_n \beta$ with $\tilde{a}$, we have

$$\begin{aligned} L_{\ell oc}(\alpha_n \beta u) &\equiv \left(\partial_t - \tilde{a}(x,t;D)\right)(\alpha_n \beta u) = f_n \\ &\text{with } f_n = [\alpha_n \beta,\ \tilde{a}]\, u\ . \end{aligned} \tag{14}$$

(2nd step.) Energy inequality.

We consider the solution $\alpha_n \beta u(\cdot,t)$ of (14). Denoting $\beta u(\cdot,t)$ by $v(\cdot,t)$,

$$\frac{d}{dt}\| \alpha_n v(\cdot,t) \|^2 = 2\,\mathrm{Re}\left(\tilde{a}(\alpha_n v),\ \alpha_n v\right) + 2\,\mathrm{Re}\,(f_n,\ \alpha_n v).$$

By (8), for large n, we have

$$\frac{d}{dt}\| \alpha_n v\| \geq \delta' n^m \| \alpha_n v\| - \| f_n \| , \tag{15}$$

δ' being a positive constant.

(3rd step.) Asymptotic expression of f_n.

$f_n = \alpha_n \beta \tilde{a} - \tilde{a} \alpha_n \beta$. Hereafter we write a instead of $\tilde{a}$.

First, by 4) in last section, we have

$$\beta a = \sum_{|\nu| \leq N} (-1)^{|\nu|} \nu!^{-1} a^{(\nu)} \beta_{(\nu)} + r_N, \qquad \text{with } r_N \in S^{m-N-1}_{1,0}\ .$$

Next,

$$\alpha_n \beta a = \sum_{|\nu| \le N} (-1)^{|\nu|} \nu!^{-1} \alpha_n a^{(\nu)} \beta_{(\nu)} + \alpha_n r_N .$$

Now we commute α_n with $a^{(\nu)}$ for each term in the above sum. By (11) (with $p = 0$, $a = a^{(\nu)}$, $N = N - |\nu|$), and observing $a^{(\nu)} \in S^{m-|\nu|}_{1,0}$, we have

$$\alpha_n a^{(\nu)} = \sum_{|\mu+\nu| \le N} \mu!^{-1} a^{(\nu)}_{(\mu)} \alpha_n^{(\mu)} + r_{\nu,N} ,$$

where $\| r_{\nu,N} \| \le \text{const. } n^{m-N-1}$, $\| \alpha_n r_N \| \le \text{const. } n^{m-N-1}$. The last inequality is derived from (12) (with $p = 0$). Thus denoting

$$S_N = \alpha_n r_N + \sum_{\nu} r_{\nu,N} \beta_{(\nu)},$$

we have

$$f_n = \sum_{1 \le |\mu+\nu| \le N} \mu!^{-1} \nu!^{-1} (-1)^{|\nu|} a^{(\nu)}_{(\mu)} \left(\alpha_n^{(\mu)} \beta_{(\nu)} u\right) + S_N u \qquad (16)$$

$$\text{with} \quad \| S_N \| \le \text{const. } n^{m-N-1} .$$

The integer N is fixed later. Keeping in mind the energy inequality (15), we estimate $\| f_n \|$. Observe $a^{(\nu)}_{(\mu)} \in S^{m-|\nu|}_{1,0}$, then applying (9), we have

$$\| f_n \| \le n^m C(N) \sum_{1 \le |\mu+\nu| \le N} n^{-|\nu|} \| \alpha_n^{(\mu)} \beta_{(\nu)} u \| + \text{const. } n^{m-N-1} \| u \| . \quad (17)$$

(4th step.) Consideration of $\alpha_n^{(p)} \beta_{(q)} u$.

The reader may be discouraged in view of the appearance of new micro-localized function $\{\alpha_n^{(\mu)} \beta_{(\nu)} u\}$, $(1 \le |\mu + \nu| \le N)$ in the (asymptotic) expression of f_n in (16). However we can argue further. The author asks reader's patience. We started from the micro-localizer $\alpha_n(D)\beta(x)$. Instead of this, if we start from the micro-localizer $\alpha_n^{(p)}\beta_{(q)}$, with $|p+q| \le N$, we shall obtain

$$L_{loc}\left(\alpha_n^{(p)}\beta_{(q)}u\right) = f_{n,p,q} \tag{18}$$

where

$$f_{n,p,q} = \sum_{1\le|\mu+\nu|\le N-|p+q|} (-1)^{|\nu|}\mu!^{-1}\nu!^{-1}a_{(\mu)}^{(\nu)}\left(\alpha_n^{(p+\mu)}\beta_{(q+\nu)}u\right) + S_{N,p,q}u$$

$$\text{with} \quad n^{-|q|}\|S_{N,p,q}\| \le \text{const. } n^{m-N-1}. \tag{19}$$

Thus we have, for $|p+q| \le N$,

$$n^{-|q|}\|f_{n,p,q}\| \le n^m C(N) \sum_{1\le|\mu+\nu|\le N-|p+q|} n^{-|q+\nu|}\|\alpha_n^{(p+\mu)}\beta_{(q+\nu)}u\|$$

$$+ \text{const. } n^{m-N-1}\|u\|. \tag{20}$$

Now (17) can be regarded as the case $p = q = 0$. Of course the energy inequality

$$\frac{d}{dt}\|\alpha_n^{(p)}\beta_{(q)}u\| \ge \delta' n^m\|\alpha_n^{(p)}\beta_{(q)}u\| - \|f_{n,p,q}\| \tag{21}$$

holds (see (15)).

Here, we should observe that, in (20), in the sum with respect to (μ, ν), the restriction $|\mu+\nu| \ge 1$, is important. So, we take M large such that

$$\sum_{|\mu+\nu|\ge 1} M^{-|\mu+\nu|} \le \tfrac{1}{2}\delta'/C(N),$$

and consider the following energy form

$$S_n(t;u) = \sum_{|p+q|\le N} M^{|p+q|_n - |q|} \| \alpha_n^{(p)} \beta_{(q)} u(\cdot,t) \| , \tag{22}$$

we have, summing up (21) with the factor $M^{|p+q|_n - |q|}$,

$$\frac{d}{dt} S_n(t;u) \ge \frac{1}{2}\delta' n^m S_n(t;u) - \text{const.}\ n^{m-N-1} \| u(\cdot,t) \| . \tag{23}$$

We take now

$$N = k + [\ m\ ] \tag{24}$$

where k is the same k appearing in (6).

($\underline{5^{th}\ \text{step}}$.)

We define a sequence of solutions $u_n(\cdot,t)$ of (4), which we supposed to exist uniquely, by the initial data $u_n(\cdot,0) \in H^\infty$, as follows: Take a function $\Psi(\xi)$ with its support in $|\xi| \le 1$, and $\int |\Psi(\xi)|^2 d\xi = 1$. Let $\Psi_n(\xi) = \Psi(\xi - n\xi^0)$, and denote its inverse Fourier image by $\tilde{\Psi}_n(x)$. Then $u_n(\cdot,0)$ is defined by

$$u_n(\cdot,0) = \tilde{\Psi}_n(x) . \tag{25}$$

We have

$$\begin{cases} \text{i)} \quad \| u_n(\cdot,0) \|_{H^k} = O(n^k) \\ \text{ii)} \quad \| \alpha_n \beta u_n(\cdot,0) \|_{L^2} \ge c_0 \quad (>0) \end{cases} \tag{26}$$

i) is derived from Plancherel theorem, in view of

$$\partial^{\alpha}\tilde{\Psi}_n = (2\pi)^{-\ell} \int e^{ix\xi}(i\xi)^{\alpha}\, \Psi(\xi - n\xi^0)\, d\xi.$$

ii) can be seen as follows. First, denoting the Fourier inverse image $\Psi(\xi)$ by $\tilde{\Psi}(x)$, we have $\tilde{\Psi}_n(x) = e^{in\xi^0 x}\,\tilde{\Psi}(x)$, and $\tilde{\Psi}(x)$ is real-analytic Next, commuting α_n with β, we have

$$\alpha_n \beta \tilde{\Psi}_n = \sum_{|\nu| \leq N} \nu!^{-1} \beta_{(\nu)} \alpha_n^{(\nu)} \tilde{\Psi}_n + r_N \tilde{\Psi}_n.$$

Since $\alpha_n^{(\nu)} \tilde{\Psi}_n = 0$ for $|\nu| \geq 1$, and $\alpha_n \tilde{\Psi}_n = \tilde{\Psi}_n$, we have

$$\alpha_n \beta \tilde{\Psi}_n = \beta \tilde{\Psi}_n + r_N \tilde{\Psi}_n,$$

where we can apply (11) to r_N as a very special case (taking $a = \beta(x)$, $p = 0$), which implies $\| r_N \tilde{\Psi}_n \| \leq \text{const.}\; n^{-N-1} \| \tilde{\Psi}_n \| = O(n^{-N-1})$. On the other hand $\| \beta\tilde{\Psi}_n \| = \| \beta\Psi \| > 0$, which proves ii).

(<u>Final step</u>.)

In (23), we put $u = u_n(\cdot, t)$, which implies

$$S_n(0; u_n) \geq \| \alpha_n \beta u_n(\cdot, 0) \| \geq c_0.$$

Now, (26), i), combined with (6), implies

$$\| u_n(\cdot, t) \| = O(n^k), \qquad t \in [0, T]. \tag{27}$$

This implies, in particular

$$S_n(t; u_n) = O(n^k).$$

On the other hand, (23), combined with (24) and (27), implies

$$\frac{d}{dt} S_n(t; u_n) \geq \frac{1}{2}\delta' \, n^m \, S_n(t; u_n) - o(1).$$

By integration, we get

$$S_n(t; u_n) \geq \exp\left(\frac{1}{2}\delta' \, n^m t\right) \left[S_n(0; u_n) - o(1) \right]$$

$$\geq \exp\left(\frac{1}{2}\delta' \, n^m t\right) c_0/2 \tag{28}$$

for n large. (27) and (28) are not compatible if n is large and $t > 0$, which proves Theorem 3.

§ 4. What about the proof of Theorem 2 ? We explain this briefly, because the principle of proof becomes clear by the detailed argument of § 3. For simplicity, we consider almost equivalent problem.

Let us consider the first order system (as Courant's school prefer).

$$\begin{cases} \partial_t u - \sum_{j=1}^{\ell} A_j(x,t)\partial_{x_j} u - B(x,t)\, u = 0 \\ u\big|_{t=0} = u_0(x) \end{cases} \tag{29}$$

where $u = {}^t(u_1, \ldots, u_N)$, and $A_j(x,t)$ $(1 \leq j \leq \ell)$, $B(x,t)$ are $N \times N$ matrices. In this case the characteristic equation becomes

$$\det\left(\lambda I - \sum_j A_j \xi_j \right) = 0.$$

Denote the characteristic roots by $\lambda_i(x,t;\xi)$. Then Theorem 2 becomes

<u>Theorem 2'.</u> In order that the local Cauchy problem at the origin $(x,t) = (0,0)$ be C^∞-wellposed, the condition:

(C) $\qquad \forall i, \forall \xi \in \mathbb{R}^\ell, \quad \lambda_i(0,0;\xi)$ are real

is necessary.

To prove this, we carry out the same arguments as those in § 3, combined with those of Lecture II, Theorem 3. We describe the argument briefly. First, we use Theorem 1. For this purpose, we take a function $\zeta(x) \in C_0^\infty$, $\zeta(x) \equiv 1$ in a neighborhood of $x = 0$, and that $\mathrm{supp}[\zeta] \times [0, t_0] \subset V_0$, if t_0 is small. Then taking $\zeta(x)u(x,t)$ instead of $u(x,t)$, (29) becomes

$$\partial_t(\zeta u) - \sum_j A_j \partial_{x_j}(\zeta u) - B(\zeta u) = -\sum_j A_j \zeta_{x_j} u \equiv g(x). \tag{30}$$

Observe that, $g(x)$ has its support in the closure $\{x;\ \zeta'(x) \neq 0\}$. Thus it has positive distance from $x = 0$. So, if we choose the support of $\beta(x)$ small, $\beta(x)\zeta u = \beta u$, and $\beta g(x) = 0$. In other words, $g(x)$ has no effect in our arguments. <u>We write hereafter u instead of ζu.</u>

First, we write (29) in the form

$$\Big(\partial_t - iA(x,t;D) - B(x,t)\Big)u = 0,$$

where $A(x,t;\xi) = \sum A_j(x,t)\xi_j$. Suppose at the origin there exists an eigen-value, say $\lambda_1(x,t;\xi)$ not real. So, if necessary, by changing ξ by $-\xi$, we can assume

$$\exists\, \xi^0 \ (\in S^{n-1}), \quad \mathrm{Im}\,\lambda_1(0,0;\xi^0) < 0. \tag{31}$$

We apply Lemma in Lecture II, Appendix to the (constant) matrix $A(0,0;\ \xi^0)$, which yields

$$N_0 A(0,0;\ \xi^0) N_0^{-1} = \begin{pmatrix} \lambda_1 & & & \\ & \lambda_2 & & \text{O} \\ & & \ddots & \\ a_{ij}^* & & & \lambda_N \end{pmatrix},$$

where $|a_{ij}^*|$ can be made small as we wish, for instance less than $\frac{\varepsilon}{2}$, and $\lambda_i = \lambda_i(0,0;\ \xi^0)$. This implies the following

$$N_0 A(x,t;\ \xi) = \begin{pmatrix} \lambda_1 & & & \\ & \lambda_2 & & \text{O} \\ & & \ddots & \\ \text{O} & & & \lambda_N \end{pmatrix} |\xi| N_0 + \left(\tilde{a}_{ij}(x,t;\ \xi) \right) N_0,$$

where $|a_{ij}(x,t;\ \xi)| < \varepsilon|\xi|$ when (x,t) is restricted to a small neighborhood of $(x,t) = (0,0)$, and ξ to a small conic neighborhood V_{ξ^0} of ξ^0. Observe again λ_i ($= \lambda_i(0,0;\ \xi^0)$) are constants.

So, the equation (30) becomes

$$\partial_t(N_0 u) = iD_0 \Lambda u + \left(A_\varepsilon(x,t;D) + B'(x,t)\right) N_0 u + N_0 g \tag{32}$$

where $A_\varepsilon(x,t;\ \xi) = \left(a_{ij}(x,t;\ \xi)\right)$, and $B' = N_0 B(x,t) N_0^{-1}$, and $\widehat{\Lambda u} = |\xi|\hat{u}(\xi)$. We remark here three things.

i) We operate to (32) the micro-localizer $\alpha_n^{(p)}\beta_{(q)}$. First, $\alpha_n^{(p)}\beta_{(q)} g = 0$. Next, in the consideration of $[\alpha_n^{(p)}\beta_{(q)}, \Lambda]$, the symbol $|\xi|$ of Λ is not C^∞, namely the differentiability is violated at the origin. But this is not at all serious. We decompose $|\xi|$, by taking $\zeta(\xi) \in C_0^{\infty}$,

$$|\xi| = \zeta(\xi)|\xi| + \big(1 - \zeta(\xi)\big)|\xi|,$$

where $\zeta(\xi) = 1$ for $|\xi| \leq 1$, $= 0$ for $|\xi| \geq 2$. We express this decomposition by $\Lambda = \Lambda_0 + \Lambda_1$. Λ_1 is essentially equal to $|\xi|$. Now it is easy to see that $\| \alpha_n^{(p)} \beta_{(q)} \Lambda_0 \|$ is estimated by any negative power of n when $n \to \infty$, by considering the commutator $[\alpha_n^{(p)}, \beta_{(q)}]\Lambda_0$. Concerning $[\alpha_n^{(p)}\beta_{(q)}, \Lambda_1]$, its asymptotic expression is entirely the same as when we suppose $\Lambda_1 = \Lambda$.

ii) In order to estimate $a_{ij}(x,t;D)\alpha_n^{(p)}\beta_{(q)}u$, we use the following statement of sharp Gårding inequality. Let $m > 0$, $a(x, \xi) \in S^m_{1,0}$, and $a(x, \xi) = a_m(x, \xi) + b(x, \xi)$ (order $b < m$). a_m is essentially homogeneous of degree m. Then

$$\| a(x,D)\alpha_n v\| \leq |a_m|_0 \{(1 + \varepsilon')n\}^m \| \alpha_n v\| + C\, n^{m-\frac{1}{2}} \| \alpha_n v\| ,$$

where $|a_m|_0 = \sup|a_m(x, \xi')|$, sup is taken over $(x, \xi) \in \mathbb{R}^{\ell} \times V_{\xi^0}$, and $\xi' = \xi/|\xi|$. Let us observe that, as we observe in the proof of Theorem 3, $|a_{ij}(x,t;\xi)|_0$ can be supposed $\leq \varepsilon$.

iii) In the proof of Lecture II, Theorem 3, we used S_n by using $v_i(t;\xi)$. Here we use the similar form (compare with (22)): Without loss of generality we assume

$$\operatorname{Re} i\lambda_1, \operatorname{Re} i\lambda_2, \cdots, \operatorname{Re} i\lambda_{k_0} > 0,$$

$$\operatorname{Re} i\lambda_{k_0+1}, \cdots, \operatorname{Re} i\lambda_N \leq 0.$$

Let $\min_{1\le i\le k_0} \operatorname{Re} i\lambda_i = \delta''$. Denoting $V = N_0 u = (v_1(\cdot,t), \ldots, v_N(\cdot,t))$, we define

$$S_n(t;u) = \sum_{|p+q|\le N} M^{|p+q|} n^{-|q|} \exp(-\delta'' nt) \left[\sum_i' \| \alpha_n{}^{(p)} \beta_{(q)} v_i(\cdot,t) \|^2 - \sum_i'' \| \alpha_n{}^{(p)} \beta_{(q)} v_i(\cdot,t) \|^2 \right],$$

where $\sum_i'$ is taken over $1 \le i \le k_0$, $\sum_i''$ over $k_0 + 1 \le i \le N$. The detailed argument is left to the reader (see, Mizohata [7]).

§ 5. Further considerations. Theorem 3 can be extended to more general cases. Let us consider

$$\partial_t^m u + \sum_{j=1}^m a_j(x,t;D)\partial_t^{m-j} u = f \tag{33}$$

where $a_j(x,t;\xi) \in S^{pj}_{1,0}$ $(1 \le j \le m)$ with $p > 0$ (p may be any positive number), depending smoothly on t. Let the characteristic roots of

$$\lambda^m + \sum_j \mathring{a}_j(x,t;\xi)\lambda^{m-j} = 0$$

be $\lambda_i(x,t;\xi)$, where $\mathring{a}_j(x,t;\xi)$ is the homogeneous part of degree pj of a_j. Then,

Theorem 4. In order that the forward Cauchy problem of (33) in $t \in [0, T]$ be uniformly H^∞-wellposed, it is necessary (but not sufficient in general) that the following condition is satisfied:

$$\forall\, i, \quad \forall\, (x,t;\xi) \in \mathbb{R}^\ell \times [0, T] \times \mathbb{R}^\ell, \quad \operatorname{Re} \lambda_i(x,t;\xi) \le 0.$$

Finally we discuss the H^∞-wellposedness for the elementary operator

$$\begin{cases} \left(\partial_t - i\lambda(x,t;D) - c(x,t;D)\right)u(x,t) = f \\ \\ u\Big|_{t=0} = u_0(x) \in H^\infty, \end{cases} \tag{34}$$

defined in $(x,t) \in \mathbb{R}^\ell \times [0, T]$, where

i) $\lambda(x,t;\xi) \in S^1_{1,0}$, essentially homogeneous of degree 1 in ξ and <u>real-valued</u>,

ii) $c(x,t;\xi) \in S^\rho_{1,0}$, with $\rho < 1$, and it can be written as $c(x,t;\xi) = \mathring{c}(x,t;\xi) + \tilde{c}(x,t;\xi)$, where $\mathring{c}$ is essentially homogeneous of degree ρ, and order $\tilde{c} \le \rho - 1$.

Then

<u>Theorem 5</u>. (<u>Levi condition</u>) In order that the forward Cauchy problem for (32) be uniformly H^∞-wellposed in $t \in [0,T]$, it is necessary and sufficient that, either $\rho \le 0$ or if $\rho > 0$, we have

$$\forall (x, \xi, t), \quad \text{Re}\ \mathring{c}(x,t;\xi) \le 0.$$

<u>Proof</u>. The sufficiency is almost evident. We show its necessity. Its proof is essentially the same as that of Theorem 3. We explain only the main points. Suppose the condition is violated. Then, without loss of generality, we can assume

$$\exists \xi^0, \ |\xi^0| = 1, \quad \text{such that} \quad \text{Re}\ \mathring{c}(0,0;\xi^0) > 0 . \tag{35}$$

We assume moreover (34) is H^∞-wellposed for $t \in [0, T]$. From these two assumptions we show a contradiction. First,for micro-localized solution around $(x, \xi) = (0, \xi^0)$, we have the following energy inequality

$$\frac{d}{dt}\| \alpha_n^{(p)}\beta_{(q)}u(\cdot,t)\| \geq \delta n^\rho \| \alpha_n^{(p)}\beta_{(q)}u(\cdot,t)\| - \| f_{n,p,q}\| \quad (36)$$

where $0 < \rho < 1$, and $\delta > 0$ (compare with (21)).

To fix the idea, consider the situation with $\alpha_n\beta$ u. Putting $a = i\lambda + c$, we get

$$(\partial_t - \tilde{a})(\alpha_n \beta u) = f_n .$$

Here

$$f_n \sim \sum_{|\mu+\nu|\geq 1} (-1)^{|\nu|}\mu!^{-1}\nu!^{-1}\tilde{a}^{(\nu)}_{(\mu)}\left(\alpha_n^{(\mu)}\beta_{(\nu)}u\right)$$

where $\tilde{a}$ is a modification of a before the commutation with β(see § 3).

$$\| \tilde{a}^{(\nu)}_{(\mu)}\| \leq \text{const. } n^{1-|\nu|}.$$

The term with $|\mu + \nu| = 1$ are decisive terms.

$$\| \tilde{a}^{(e_j)}(\alpha_n\beta_{(e_j)} u)\| \leq \text{const.}\| \alpha_n\beta_{(e_j)} u\| ,$$

$$\| \tilde{a}_{(e_j)}(\alpha_n^{(e_j)} \beta u)\| \leq \text{const. } n\| \alpha_n^{(e_j)} \beta u\| .$$

Compare these terms with (36) for $|p + q| = 1$.It would be natural to consider $n^{-\rho}\| \alpha_n\beta_{(e_j)} u\|$, $n^{1-\rho}\| \alpha_n^{(e_j)}\beta u\|$. In general we consider

$$n^{(1-\rho)|p|-\rho|q|}\| \alpha_n^{(p)}\beta_{(q)} u\| .$$

In fact, in the expression

$$f_{n,p,q} \sim \sum_{|\mu+\nu|\geq 1} (-1)^{|\nu|}\mu!^{-1}\nu!^{-1}\tilde{a}^{(\nu)}_{(\mu)}\left(\alpha_n^{(p+\mu)}\beta_{(q+\nu)}\right),$$

$n^{(1-\rho)|p|-\rho|q|}\|\tilde{a}^{(\nu)}_{(\mu)}\alpha_n^{(p+\mu)}\beta_{(q+\nu)}u\|$ is estimated in the form,

$$\text{const. } n^{\rho}\left\{n^{-(1-\rho)(|\mu+\nu|-1)}\right\}n^{(1-\rho)|p+\mu|-\rho|q+\nu|}\|\alpha_n^{(p+\mu)}\beta_{(q+\nu)}u\|.$$

This shows that, under these weights, only the terms with $|\mu+\nu| = 1$ are delicate. Thus, we are led to consider

$$S_n(t;u) = \sum_{|p+q|\leq N} M^{|p+q|}n^{(1-\rho)|p|-\rho|q|}\|\alpha_n^{(p)}\beta_{(q)}u(\cdot,t)\|,$$

where $N = [{}^{k}/_{\rho}] + 1$ (compare with (24)). The detailed argument is left to the reader.

Appendix

§ A.1. Preliminaries.

In this appendix, we give the proof of several facts concerning the elementary properties of pseudo-differential operators. First, we explain some basic facts. For details, see for instance Kumano-go [5].

Let $a(x, \xi) \in S^m_{1,0}$. This means $a(x, \xi) \in C^\infty(\mathbb{R}^\ell_n \times \mathbb{R}^\ell_\xi)$, and that

$$\forall (\alpha, \beta) \qquad |a^{(\alpha)}_{(\beta)}(x, \xi)| \le C_{\alpha\beta}(1 + |\xi|)^{m-|\alpha|},$$

where $a^{(\alpha)}_{(\beta)} = \partial^\alpha_\xi(i^{-1}\partial_x)^\beta a(x, \xi) \quad (= \partial^\alpha_\xi D_x^\beta a(x, \xi)).$

Theorem A.1. (Calderon-Vaillancourt)

Let $a(x, \xi) \in S^0_{0,0}$. Then

$$\| a(x,D) \| \le c(\ell) \sum_{\substack{|\alpha| \le \ell+2 \\ |\beta| \le \ell+2}} \sup |a^{(\alpha)}_{(\beta)}(x, \xi)|,$$

where sup is taken over $\mathbb{R}^\ell_x \times \mathbb{R}^\ell_\xi$, and $\| \cdot \|$ means the L^2-(operator) norm.

For our arguments, the estimates of remainder terms of some asymptotic expansion are important. This is always deduced from the Taylor expansion:

$$f(x + y) = \sum_{|\nu| \le N} \nu!^{-1}\partial^\nu f(x)y^\nu + (N+1)\int_0^1 (1 - \theta)^N f_{N,\theta}(x,y)\, d\theta \tag{A.1}$$

where $f_{N,\theta} = \sum_{|\nu|=N+1} \nu!^{-1}\partial^\nu f(x+\theta y)\, y^\nu$.

We use after the following expression. Let $a(x, \xi) \in S^m_{1,0}$, $b(x, \xi) \in S^{m'}_{1,0}$.

Then $a(x,D)\circ b(x,D)$ is also a pseudo-differential operator whose symbol is expressed by

$$c(x, \xi) = (2\pi)^{-\ell}\, \mathrm{Os}\text{-}\iint e^{-iy\eta} a(x, \xi+\eta) b(x+y, \xi)\, dy d\eta. \tag{A.2}$$

Hereafter we omit the notation Os- (= oscillatory integral).

We use several times cut-off functions. We fix here a standard one: $\chi(\xi) \in C_0^\infty$, $= 1$ for $|\xi| \le \frac{1}{4}$, $= 0$ for $|\xi| \ge \frac{1}{2}$, $0 \le \chi(\xi) \le 1$.

§ A.2. Proof of (12).

The symbol of $a(x,D)\beta(x)$ is expressed by (A.2),

$$(2\pi)^{-\ell} \iint e^{-iy\eta} a(x, \xi+\eta)\beta(x+y)\, dy d\eta. \tag{A.3}$$

That of $\beta(x)a(x,D)$ is $\beta(x)a(x, \xi)$, which can be written by

$$(2\pi)^{-\ell} \iint e^{-iy\eta} \beta(x) a(x, \xi+\eta)\, dy d\eta.$$

Now, applying (A.1) to $\beta(x)$, we have

$$\beta(x) = \beta(x+y-y) = \sum_{|\nu|\le N} (-1)^{|\nu|} \nu!^{-1} \beta_{(\nu)}(x+y)(iy)^\nu + (N+1)\int_0^1 (1-\theta)^N \beta_{N,\theta}(x,y)\, d\theta,$$

$$\beta_{N,\theta} = \sum_{|\nu|=N+1} (-1)^{N+1} \nu!^{-1} \beta_{(\nu)}(x+y-\theta y)(iy)^\nu.$$

In view of the form of (A.3), by integration by parts we have

$$\beta(x)a(x,D) = \sum_{|\nu|\leq N} (-1)^{|\nu|}\nu!^{-1}a^{(\nu)}(x,D)\beta_{(\nu)} + r_N(x,D) \tag{A.4}$$

with
$$r_N(x,\xi) = (N+1)\int_0^1 (1-\theta)^N r_{N,\theta}(x,\xi)\,d\theta,$$

$$r_{N,\theta}(x,\xi) = \sum_{|\nu|=N+1} (-1)^{N+1}\nu!^{-1}(2\pi)^{-\ell}\iint e^{-iy\eta}a^{(\nu)}(x,\xi+\eta)\beta_{(\nu)}(x+y-\theta y)\,dyd\eta.$$

We consider one of the last integrals. The decomposition by using $\chi(\eta)$ in § A.1 gives

$$\iint e^{-iy\eta}a^{(\nu)}(x,\xi+\eta)\chi(\tfrac{\eta}{|\xi|})\beta_{(\nu)}(x+y-\theta y)\,dyd\eta$$

$$+ \iint e^{-iy\eta}a^{(\nu)}(x,\xi+\eta)\left[1-\chi(\tfrac{\eta}{|\xi|})\right]\beta_{(\nu)}(x+y-\theta y)\,dyd\eta$$

$$= I_1 + I_2.$$

These integrals are, in general, oscillatory integrals, and in order to estimate $I_1(x,\xi)$, independent of θ, we express this as the usual integral by integration by parts,

$$I_1(x,\xi;\theta) = \iint e^{-iy\eta}(1-\Delta_\eta)^\ell\left[a^{(\nu)}(x,\xi+\eta)\chi(\tfrac{\eta}{|\xi|})\right](1+|\eta|^2)^{-\ell}\cdot$$

$$(1-\Delta_y)^\ell\left[\beta_{(\nu)}(x+y-\theta y)/(1+|y|^2)^\ell\right]\,dyd\eta.$$

Now, $$\left|(1-\Delta_\eta)^\ell\left[a^{(\nu)}(x,\xi+\eta)\chi(\tfrac{\eta}{|\xi|})\right]\right| \leq \text{const.}(1+|\xi|)^{m-N-1},$$

where const. is independent of ξ, η. So that,

$$|I_1(x,\xi;\theta)| \leq \text{const.}(1+|\xi|)^{m-N-1},$$

where const. is also independent of θ. We have the same type inequality for $\partial_\xi^p \partial_x^q I_1(x,\xi;\theta)$:

$$|\partial_\xi^p \partial_x^q I_1(x,\xi;\theta)| \leq \text{const.}(1 + |\xi|)^{m-N-1-|p|}.$$

Thus, we have

$$\int_0^1 (1 - \theta)^N I_1(x,\xi;\theta)\, d\theta \in S_{1,0}^{m-N-1}. \tag{A.5}$$

Concerning $I_2(x,\xi;\theta)$, by taking M large,

$$I_2 = \iint e^{-iy\eta}(1-\Delta_\eta)^\ell \left[\frac{a^{(\nu)}(x,\xi+\eta)}{|\eta|^{2M}}\left(1-\chi\left(\frac{\eta}{|\xi|}\right)\right)\right](-\Delta_y)^M \beta_{(\nu)}(x+y-\theta y)\Big/(1+|y|^2)^\ell\, dy d\eta.$$

It is easy to see that (A.5) is also true for $I_2(x,\xi;\theta)$. Thus we proved, in the expression (A.4),

$$r_N(x, \xi) \in S_{1,0}^{m-N-1}.$$

§ A.3. Partition of unity.

The proofs of (10), (11) are delicate. For this purpose, we prepare the following basic partition of unity in the ξ-space. We define $\phi_k(\rho)$, $\rho \in \mathbf{R}_+$, in the following way.

$$0 \leq \phi_k(\rho) \leq 1, \quad \phi_k(\rho) \in C_0^\infty, \quad \operatorname{supp}[\phi_k(\rho)] \subset [\tfrac{1}{3}4^k, 3\cdot 4^k],$$

$$\sum_{k=0}^\infty \phi_k(\rho) = 1; \quad |\partial^m \phi_k(\rho)| \leq \frac{c_m}{R_k^m} \quad (m = 0, 1, 2, \cdots)\ (R_k = 4^k).$$

In accordance with parameter n, we define

$\chi_n(\xi) = \sum_j \phi_j(|\xi|)$, where j runs through such ϕ_j that

$$\mathrm{supp}[\phi_j] \cap \mathrm{supp}[\alpha_n] \neq \phi,$$

$\chi_{n,-}(\xi) = \sum_j \phi_j(|\xi|)$, where j runs through such ϕ_j that

$$\mathrm{supp}[\phi_j] \subset \{\xi ; |\xi| < n\} \quad \text{and} \quad \mathrm{supp}[\phi_j] \cap \mathrm{supp}[\alpha_n] = \phi,$$

$\chi_{n,+}(\xi)$ is defined in the same way as above by replacing $\{\xi ; |\xi| < n\}$ by $\{\xi ; |\xi| > n\}$. We have

$$\chi_n(\xi) + \chi_{n,-}(\xi) + \chi_{n,+}(\xi) = 1.$$

Moreover

$$|\partial^\alpha \chi_n(\xi)|, \quad |\partial^\alpha \chi_{n,\pm}(\xi)| \le \frac{c_\alpha}{n^{|\alpha|}}. \tag{A.7}$$

§ A.4. Estimates of $\alpha_n(D)b(x,D)\chi_{n,\pm}(D)$.

For convenience, we introduce the word " negligible ". We say that an operator depending on parameter n, $a_n \in (L^2, L^2)$, or $f_n \in L^2$, is <u>negligible</u> if $\|a_n\|$ or $\|f_n\|$ is estimated by any negative power of n, when $n \to \infty$.

<u>Proposition A.1</u>. Let $b(x, \xi) \in S^m_{1,0}$ (m is any real number). Then

$$\alpha_n(D)b(x,D)\chi_{n,\pm}(D) \quad \text{are negligible.}$$

<u>Proof.</u> Consider first $\alpha_n b \chi_{n,-}$. We take a suitable $\tilde{\chi}_n(\xi)$, with $\tilde{\chi}_n = 1$, on the support of $\chi_{n,-}(\xi)$, and with the estimates (A.7). Evidently

$\chi_{n,-} = b\tilde{\chi}_n \chi_{n,-}$. So we denote $b\tilde{\chi}_n = b_n$.

Now we commute $\chi_{n,-}$ with b_n.

$$\alpha_n b\chi_{n,-} = \alpha_n b_n \chi_{n,-} = \sum_{|\nu|\leq N'} \alpha_n (-1)^{|\nu|} \nu!^{-1} \chi_{n,-}^{(\nu)} b_{n(\nu)} + \alpha_n S_{N'},$$

where

$$S_{N'} = (N' + 1) \int_0^1 (1 - \theta)^{N'} S_{N',\theta}(x, \xi)\, d\theta,$$

$$S_{N',\theta} = \sum_{|\nu|=N'+1} (-1)^{N'+1} \nu!^{-1} (2\pi)^{-\ell} \iint e^{-iy\eta} b_{n(\nu)}(x+y,\xi) \chi_{n,-}^{(\nu)}(\xi+\eta-\theta\eta)\, dyd\eta.$$

First $\alpha_n \chi_{n,-}^{(\nu)} = 0$. Thus $\alpha_n b\chi_{n,-} = \alpha_n S_{N'}$. Next, as in § A.2, the integral in above expression can be expressed by integration by parts,

$$\iint e^{-iy\eta}(1-\Delta_y)^{\ell}\left[b_{n(\nu)}(x+y,\xi)\left(1+|y|^2\right)^{-\ell}\right]\left(1+|\eta|^2\right)^{-\ell}\left[(1-\Delta_\eta)^{\ell}\chi_{n,-}^{(\nu)}(\xi+\eta-\theta\eta)\right]dyd\eta.$$

This integral is estimated by const. $n^{m_+ - N' - 1}$ ($m_+ = \max(m, 0)$), where const. is independent of θ. This is the same with its derivatives with respect to x and ξ. Since N' can be taken arbitrarily large, this shows $\| S_{N'} \|$ is estimated by any negative power of n.

We pass to the estimate of $\alpha_n b\chi_{n,+}$. This estimate can be carried out in the following way. From the definition of $\chi_{n,+}$ ($\chi_{n,+} = \sum' \phi_k$),

$$\alpha_n b\chi_{n,+} = \sum_k{}' \alpha_n b\phi_k.$$

In the expression of the right hand side, we can assume $b(x, \xi)$ has its support in a suitable neighborhood of that of $\phi_k(|\xi|)$ as function in ξ.

We consider each term of the right hand side. In the same way as above, we commute b with ϕ_k. Thus, since $\alpha_n \phi_k^{(\nu)} = 0$, we have

$$\alpha_n b\phi_k = \alpha_n S_{k,N'} ,$$

where the symbol is the same as above replacing $\chi_{n,+}$ by ϕ_k. Thus, we have

$$\| S_{k,N'} \| \le \text{const.} / R_k^{N'+1-m_+} ,$$

where const. is independent of k. Thus

$$\sum_k{}' \| S_{k,N'} \| \le \text{const.} / n^{N'+1-m_+} .$$

Since N' is arbitrary, we proved that $\| \alpha_n b\chi_{n,+} \|$ is negligible.

§ A.5. Proof of (9), (10), (11).

First we prove (9). This is fairly easy. First, let $\tilde{\alpha}(\xi)$ be of small support around ξ^0 ($|\xi^0| = 1$), such that on the support of $\alpha(\xi), \tilde{\alpha}(\xi) = 1$. Put $\tilde{\alpha}_n(\xi) = \tilde{\alpha}(\xi/n)$. Then $a(x,D)\alpha_n = a(x,D)\tilde{\alpha}_n \alpha_n$. Now we apply Theorem A.1 to $a(x,D)\tilde{\alpha}_n(D)$, observing that its symbol is $a(x,\xi)\,\tilde{\alpha}_n(\xi)$.

Now we prove (10), (11). First, by Proposition A.1, if we decompose

$$\| \alpha_n av \| \le \| \alpha_n a\chi_n v \| + \| \alpha_n a\chi_{n,-} v \| + \| \alpha_n a\chi_{n,+} v \| ,$$

we know that the second and third terms are negligible. So, it is enough to consider $\|\alpha_n a\chi_n v\|$. Put $a\chi_n$ (whose symbol is $a(x,\xi)\chi_n(\xi)$) $= a_n(x,D)$. Then

$$\alpha_n a_n v = \sum_{|\nu|\le N} \nu!^{-1} a_{n(\nu)}\alpha_n^{(\nu)} v + r_N(x,D)v \tag{A.8}$$

with

$$r_N(x,\xi) = (N+1)\int_0^1 (1-\theta)^N r_{N,\theta}(x,\xi)\, d\theta,$$

$$r_{N,\theta} = \sum_{|\nu|=N+1} \nu!^{-1}(2\pi)^{-\ell}\iint e^{-iy\eta}\alpha_n^{(\nu)}(\xi+\theta\eta)a_{n(\nu)}(x+y,\xi)\, dyd\eta.$$

First, since $a_{n(\nu)}(x+y,\xi) = a_{(\nu)}(x+y,\xi)\chi_n(\xi)$, by using the same argument as we used several times, we see easily

$$\| r_N(x,D)\| \le \text{const. } n^{m-N-1}.$$

Next, in the expression (A.8), the symbol of $a_{n(\nu)}(x,D)\alpha_n^{(\nu)}(D)$ is $a_{(\nu)}(x,\xi)\chi_n(\xi)\alpha_n^{(\nu)}(\xi)$. Since, on the support of $\alpha_n^{(\nu)}(\xi)$, $\chi_n(\xi) = 1$, this symbol is equal to $a_{(\nu)}(x,\xi)\alpha_n^{(\nu)}(\xi)$. Thus (A.8) becomes

$$\alpha_n a_n v = \sum_{|\nu|\le N} \nu!^{-1} a_{(\nu)}\alpha_n^{(\nu)} v + r_N v.$$

Evidently, by (9), $\| a_{(\nu)}\alpha_n^{(\nu)}v\| \le \text{const. } n^m\|\alpha_n^{(\nu)}v\| \le \text{const. } n^{m-|\nu|}\|v\|$. This proves (10) and (11). Finally $(10)_p$ is evident from the above proof.

§ A.6. In our argument, the commutator $[\alpha_n(D)\beta(x), a(x,D)]$ plays a crucial role. We give here a comment to its asymptotic expansion.

First

$$\alpha_n(D)\beta(x) \longmapsto (2\pi)^{-\ell}\iint e^{-iy\eta}\alpha_n(\xi+\eta)\beta(x+y)\, dyd\eta.$$

This means that the symbol of $\alpha_n(D)\beta(x)$ is expressed by the integral. Applying (A.2),

$$\alpha_n(D)\beta(x)a(x,D) \mapsto (2\pi)^{-2\ell}\iiiint e^{-iy\eta-iy'\eta'}\alpha_n(\xi+\eta+\eta')\beta(x+y)a(x+y',\xi)\, dyd\eta dy'd\eta'.$$

In the same way,

$$a(x,D)\alpha_n(D)\beta(x) \mapsto (2\pi)^{-2\ell}\iiiint e^{-iy\eta-iy'\eta'}\alpha_n(\xi+\eta+\eta')\beta(x+y)a(x,\xi+\eta)\, dyd\eta dy'd\eta'.$$

Therefore, in order to consider the difference of the above symbols, we are led to consider the difference $a(x+y',\xi) - a(x,\xi+\eta)$. By Taylor expansion,

$$a(x+y',\xi) - a(x,\xi+\eta) = \sum_{1\le|\mu+\nu|\le N} \mu!^{-1}\nu!^{-1}\partial_x^{\mu}\partial_\xi^{\nu}a(x,\xi+\eta)y'^{\mu}(-1)^{|\nu|}\eta^{\nu}$$

$$+ (N+1)\int_0^1 (1-\theta)^N s_{N,\theta}(u,\xi)\, d\theta,$$

where

$$s_{N,\theta} = \sum_{|\mu+\nu|=N+1} \mu!^{-1}\nu!^{-1}(-1)^{|\nu|}\partial_x^{\mu}\partial_\xi^{\nu}a(x+\theta y',\xi+\eta-\theta\eta)y'^{\mu}\eta^{\nu}.$$

Thus,

$$\alpha_n(D)\beta(x)a(x,D) - a(x,D)\alpha_n(D)\beta(x) = \sum_{1\le|\mu+\nu|\le N} \mu!^{-1}\nu!^{-1}(-1)^{|\nu|}a^{(\nu)}_{(\mu)}\alpha_n^{(\mu)}\beta_{(\nu)}$$

$$+ r_N(x,D),$$

where

$$r_N(x, \xi) = (N+1)\int_0^1 (1-\theta)^N r_{N,\theta}(x, \xi)\, d\theta,$$

$$r_{N,\theta} = \sum_{|\mu+\nu|=N+1} (-1)^{|\nu|}\mu!^{-1}\nu!^{-1}(2\pi)^{-2\ell}\iiiint e^{-iy\eta-iy'\eta'}\alpha_n^{(\mu)}(\xi+\eta+\eta')\beta_{(\nu)}(x+y)$$

$$\times\, a_{(\mu)}^{(\nu)}(x+\theta y', \xi+\eta-\theta\eta)\; dyd\eta dy'd\eta'.$$

The expression of $r_N(x, \xi)$ shows the necessity of introducing the partition of unity in the ξ-space, because we can not say anything about

$$a_{(\mu)}^{(\nu)}(x + \theta y', \xi + \eta - \theta\eta).$$

If we consider $a(x, \xi)\chi_n(\xi)$ instead of $a(x, \xi)$ itself, we know that

$$\left|\left(a(x, \xi)\chi_n(\xi)\right)_{(\mu)}^{(\nu)}\right| \le \text{const. } n^{m-|\nu|}.$$

In this case,we obtain the desired estimate

$$\| r_N(x,D)\| \le \text{const. } n^{m-N-1}.$$

§ A.7. The pseudo-differential operators originats from the singular integral operators of Calderón and Zygmund [1]. For simplicity, we consider the symbol $a(x, \xi)$, which is homogeneous of degree 0 in ξ. This symbol plays a central role in the treatment of partial differential operators. Following them, the operator norm in the L^2-space is estimated by

$$\| a(x,D)\| \le c(\ell)\sum_{|\alpha|\le 2\ell} \sup\left|a^{(\alpha)}(x, \xi)\right|, \tag{A.9}$$

where sup is taken over $x \in \mathbb{R}^\ell$ and $|\xi| \ge 1$.

Observe that $a(x, \xi)$ is not C^∞ in ξ. Except the trivial case, the differentiability is lost at $\xi = 0$. Even now, this result is very useful. The symbol class $S^m_{\rho,\delta}$ was introduced by Hörmander [2]. This result made the pseudo-differential operator flexible tools. He proved that if $\rho > \delta$, $a(x,D)$ becomes a bounded operator in L^2-space. Finally, Theorem A.1 shows, even in the case $S^0_{0,0}$, $a(x,D)$ is a bounded operator. Observe that $e^{ih\xi}$ (h real) belongs to this class, but not to the former class. For the proof of Theorem A.1, see for instance [5].

References

[1] A.P. Calderón and A.Zygmund, Singular integral operators and differential equations, Amer. J. Math., 79(1957), 901-21.

[2] L. Hörmander, Pseudo-differential operators and hypoelliptic equations, Proc. Symposium on Singular Integrals, Amer. Math. Soc., 10(1967), 138-83.

[3] ——————, The Cauchy problem for differential equations with double characteristics, J. Analyse Math., 32(1977), 118-196.

[4] V.Ja. Ivrii and V.M. Petkov, Necessary conditions for the Cauchy problem for non-strictly hyperbolic equations to be well-posed, Uspehi Math. Nauk, 29(1974), 3-70.

[5] H. Kumano-go, Pseudo-differential operators, M.I.T. Press, 1981.

[6] P.D. Lax, Asymptotic solutions of oscillatory initial value problem, Duke Math. J., 24(1957), 627-46.

[7] S. Mizohata, Some remarks on the Cauchy problem, J. Math. Kyoto Univ., 1(1961), 109-27.

Lecture IV. Cauchy Problems in Gevrey Class

§ 1. Introduction and results.

A C^∞-function $f(x)$ belongs to $\gamma^{(s)}$ (Gevrey class s) if $f(x)$ satisfies

$$|\partial^\alpha f(x)| \le A\alpha!^s C^{|\alpha|}, \quad {}^\forall\alpha \ge 0 \quad ({}^\exists A, C > 0).$$

It is natural to define $\gamma^{(s)}$ by saying that $f(x)$ satisfies the above condition for every compact set K, where A and C may depend on K. However, in this lecture we define as above.

Historically, Gevrey studied the properties of the solutions of the (second order) parabolic equations fairly systematically, and considered the above class of functions. For this reason, we call the above class of functions by his name. The most famous function is

$$E(x,t) = \begin{cases} \dfrac{1}{2\sqrt{\pi t}} e^{-\frac{x^2}{4t}}, & t > 0 \\ 0, & t \le 0. \end{cases}$$

This is the fundamental solution of the heat equation

$$\frac{\partial u}{\partial t} - \frac{\partial^2 u}{\partial x^2} = 0 \tag{1}$$

$E(x,t)$, for $x \ne 0$, as a function of (x,t), belongs to $\gamma^{(s)}$. Holmgren and other several mathematicians made researches related to this class. One can find these in the book by Hadamard [7]. Hörmander also touched upon this problem in his text book [9].

The simplest, however fundamental, problem is to consider the Cauchy problem for (1), imposing the Cauchy data at $x = 0$.

To consider this, it will be convenient to change x and t in (1), and write this as

$$\begin{cases} \partial_t^2 u - \partial_x u = 0 \\ u(x,0) = u_0(x) \\ \partial_t u(x,0) = u_1(x) . \end{cases} \tag{2}$$

As we shall see easily from Theorem 2 in Lecture II, <u>the Cauchy problem is not H^∞-wellposed</u>. In fact, first observe that (2) has the double characteristic root $\lambda = 0$. This satisfies the necessary condition of Lax-Mizohata theorem. However, $v(t;\xi) = e^{ix\xi + \sqrt{i}\xi t}$ shows that the condition in Lecture II, Theorem 2 fails to hold. We know that the Cauchy problem (2) is uniquely solvable for any $(u_0(x), u_1(x)) \in \gamma^{(2)}$. This could be said that <u>(2) is $\gamma^{(2)}$-wellposed</u>. We also know that this is not wellposed in $\gamma^{(s)}$ with $s > 2$. Fairly systematic treatment of the Cauchy problem in Gevrey class have been done by Leray and Ohya. Historically, in 1963, Ohya went to Paris (from Kyoto) to study the theory of P.D.E. bringing the pre-print which appeared in Journal of Math. Soc. of Japan in 1964 [22]. He met with Leray. Leray paid attention to Ohya's work, and two joint-works with Ohya appeared [16], [17]. Let us explain this more precisely. We are concerned with

$$P(u) \equiv \partial_t^m u(x,t) + \sum_{j=1}^m a_j(x,t;D_x)\partial_t^{m-j} u(x,t) = f(x,t) \tag{3}$$

where order $a_j \le j$ and that the principal symbol is of the form

$$\prod_{j=1}^k \left(\tau - \lambda_j(x,t;\xi)\right)^{m_j} . \tag{4}$$

We assume λ_j are *real and distinct* and m_j are all constant (*constant multiplicity*).

If all $m_j = 1$, we say that (3) is *strictly hyperbolic*. In this case, the C^∞-wellposedness is well-known. However if for some j, $m_j \geq 2$, (we call *non-strictly hyperbolic*, or *weakly hyperbolic*), the H^∞-wellposedness fails to hold in general. To be precise, we must impose some condition on lower order terms, which is now called *Levi condition*. However, if we assume the coefficients and the Cauchy data and f to belong to some Gevrey class, the Cauchy problem becomes uniquely solvable, *without any assumption on lower order terms*. This was proved by Ohya in the paper of 1963 mentioned above. Schematically

non-strictly hyperbolic ↗ Gevrey class

non-strictly hyperbolic ↘ Levi condition

What is the merit to treat the Cauchy problem in Gevrey class ? Contrary to the analytic functions, $\gamma^{(s)}$, for $s > 1$, contains fairly large class of C^∞-functions with compact support. (Recall that the simple C_0^∞-function $f(x) = \exp\left(-\frac{1}{1-x^2}\right)$ for $|x| < 1$, and 0 for $|x| \geq 1$, belongs to $\gamma^{(2)}$.) It is known that one of the characteristic properties of hyperbolic equations of having finite propagation property is maintained. It seems to the author that Courant school in New York takes rather negative attitude to the researches in Gevrey class. However in Europe especially in France and Soviet Union, the attitude is positive.

We explain about the admissible Gevrey class $\gamma^{(s)}$. Observe that $\gamma^{(s)} \subset \gamma^{(s')}$, if $s < s'$. Let r be the maximum multiplicity of (4): $r = \max(m_1, \cdots , m_k)$. <u>Then (3) is $\gamma^{(s)}$-wellposed if</u>

$$\frac{1}{s} > \frac{r-1}{r} \qquad \text{i.e.} \quad s < \frac{r}{r-1} \; .$$

To be precise, we assume the coefficients in (3), and f and the Cauchy data

$$\partial_t^j u \Big|_{t=0} = u_j(x) \in \gamma^{(s)} \cap C_0^\infty \quad (0 \le j \le m-1).$$

Then, there exists a unique solution $u(x,t) \in \gamma^{(s)}$ of (3).

How to understand the condition $s < \frac{r}{r-1}$? The following fact is essential to our further considerations, in particular to the micro-local analysis in Gevrey class.

Let $f(x) \in \gamma^{(s)} \cap C_0^\infty$, then

$$|\hat{f}(\xi)| \le A \exp\left(- \varepsilon |\xi|^{\frac{1}{s}}\right) \quad (\exists A, \varepsilon > o). \tag{5}$$

Conversely, if $\hat{f}(\xi) \in (\mathcal{S})$ satisfies this condition, we have $f(x) \in \gamma^{(s)}$.

For simplicity, we consider the equation

$$L(u) \equiv \left(\partial_t^m + \sum_{j=1}^{m} a_j(t;\partial_x)\, \partial_t^{m-j}\right) u = 0,$$

where <u>∂_t^m is the principal part</u>, and a_j are differential operator. Namely, we assume $\operatorname{order}_\zeta a_j(t;\zeta) \le j - 1$. Then

$$\text{order } a_j \Big/ j \leq \frac{j-1}{j} \leq \frac{m-1}{m}.$$

Let us remark that in this case m is also the multiplicity of the characteristic root $\lambda = 0$. Then, as we see easily,

$$\hat{L} v \equiv \left(\partial_t^m + \sum_j a_j(t; i\xi)\partial_t^{m-j}\right)v(t;\xi) = 0$$

has the property

$$\sum_{j=0}^{m-1} \left|\partial_t^j v(t;\xi)\right| \leq A \exp\left(\delta|\xi|^{\frac{m-1}{m}}\right). \quad \forall t \in [0, T] \ (\exists A, \delta > 0),$$

if the initial data satisfies

$$\sum_j \left|\partial_t^j v(0;\xi)\right| \leq 1.$$

This implies that, since for the initial data $(u_0(x), \cdots, u_{m-1}(x)) \in \gamma^{(s)} \cap C_0^\infty$, with $\frac{1}{s} > \frac{m-1}{m}$, we have $|\hat{u}_j(\xi)| \leq A' \exp\left(-\varepsilon|\xi|^{1/s}\right)$ $(\exists \varepsilon, A' > 0)$, by taking the fundamental system $v_0, \cdots, v_{m-1}$ of $\hat{L} v = 0$, we get the solution $u(x,t)$ by

$$\hat{u}(\xi, t) = \sum_{j=0}^{m-1} v_j(t;\xi)\, \hat{u}_j(\xi),$$

having the property $|\hat{u}(\xi,t)| \leq A'' \exp\left(-\varepsilon'|\xi|^{1/s}\right)$ $(\exists A'', \varepsilon' > 0)$.

Now we go back to the general case. What about the condition of $\lambda_j(x,t;\xi)$ in (4) ? There exists complete analogy with the C^∞ case. In fact we have (see Mizohata [19], Nishitani [20]).

Theorem 1. In order that the Cauchy problem is locally uniquely solvable at the origin for the Cauchy data in $\gamma^{(s)}$, with $s > 1$, the following cindition must hold,

$$\forall i, \xi \in \mathbb{R}^{\ell}, \quad \lambda_i(0,0;\xi) \text{ is real.}$$

Note. We assume all the coefficients of the equation (3) belong to $\gamma^{(s)}$. However in this theorem, we do not assume the constancy of the multiplicity of λ_i. In the case $s = 1$, we know by Cauchy-Kowalewski theorem, the Cauchy problem is uniquely solvable without any assumption on $\lambda_i(x,t;\xi)$. However, if we assume the following: For any initial data $\Psi \equiv (u_0, \cdots, u_{m-1}) \in C^{\omega}(\mathcal{O}_x)$, there exists a common existence domain $V_{x,t}$ of the origin such that there exists a solution $u(x,t) \in C^{\omega}(V_{x,t})$ of (3) with $f = 0$, then the condition stated in Theorem 1 should be required as a necessary condition [19], [20].

Now we go to back to (3) assuming (4). Our starting point is to use the perfect factorization of the operator (3). This simplifies remarkably the treatment of the operators with constant multiplicity. The perfect factorization means the following:

$$P = P_k \circ P_{k-1} \circ \cdots \circ P_1 + R \tag{6}$$

where the symbol $\circ$ denotes operator product, and P_j $(1 \le j \le k)$ is the pseudo-differential operator whose principal symbol is $(i\tau - i\lambda_j)^{m_j}$. To be precise,

$$P_j = (\partial_t - i\lambda_j)^{m_j} + a_{1,j}(x,t;D_x)(\partial_t - i\lambda_j)^{m_j-1} + \cdots + a_{m_j,j}(x,t;D_x) \tag{7}$$

with order $a_{ij} \leq i - 1$. We should remark here

$$(\partial_t - i\lambda_j)^m = \underbrace{(\partial_t - i\lambda_j(x,t;D_x))\circ(\partial_t - i\lambda_j)\circ \cdots \circ(\partial_t - i\lambda_j)}_{m}.$$

The terminology " perfect " is reflected in the remainder term R.

$$R = \sum_{j=1}^{m} r_j(x,t;D_x)\partial_t^{m-j},$$

where $r_j(x,t;D_x)$ are all regularizing operator, which means, usually in the class $\gamma^{(s)}$, r_j is a continuous mapping from $\mathcal{E}_x'$ to $\gamma^{(s)}$. However, in this class, since we assumed the coefficients uniform Gevrey estimates, we have (taking account of Calderón-Vaillancourt theorem),

$$\left|\partial_t^{k}\partial_x^{\alpha}\partial_\xi^{\beta} r_j(x,t;\xi)\right| \leq A\, C_1^{|\alpha|+k}(|\alpha|+k)!^{s} \exp\left(-\varepsilon_0|\xi|^{1/s}\right), \quad t \in [0, T]$$

$$(\exists A, C_1, \varepsilon_0 > 0), \quad \text{for all} \quad \alpha \geq 0,\ k \geq 0,\ |\beta| \leq n + 2.$$

Observe that we assume the coefficients Gevrey estimated in t.

Concerning this, in the case of C^∞, see Kumano-go [15], and in Gevrey class, see Nishitani [20].In the case where we do not assume Gevrey property in t on the coefficients, we shall give some comments later.

The perfect factorization is carried out, as C^∞ case, by using asymptotic expansions to obtain formal Gevrey symbols. Historically, the fundamental facts concerning Gevrey and analytic symbols have been obtained by Boutet de Monvel and Krée [1]. Recent book by Treves [24] treats this subject. The essential point in the arguments is to verify that the fundamental rules (concerning the operator products, and adjoints etc.) are maintained by establishing fairly simple correspondence from formal symbols to true symbols. For this purpose, Treves introduced pseudo-analytic symbols instead of analytic symbols. A recent work by Matsuzawa-Hashimoto-Morimoto [8] extends this method to Gevrey symbols.Incidentally, the perfect factorization in C^∞ class, is also very useful in considering the Levi condition.

Now, let us consider a factor P_j expressed by (7). Put

$$\rho_j = \max_{1\le i\le m_j} \text{order } a_{ij}(x,t;\xi) \Big/ i \quad (< 1). \tag{8}$$

Now, <u>in C^∞ case</u>, Levi condition is formulated as follows:

<u>Levi condition</u>. We say that P satisfies Levi condition if for all j,

$$\rho_j \le 0.$$

It should be remarked that we can prove that ρ_j is invariant with respect to the order of perfect factorization.

We have the following theorem, supposing all the coefficients belong to $\mathcal{B}$ (C^∞ and bounded with all their derivatives).

<u>Theorem 2</u>. In order that the forward Cauchy problem in $[0, T]$ be uniformly H^∞-wellposed, it is necessary and sufficient that P satisfies the Levi condition.

Note. Levi condition is formulated in other ways by Flaschka-Strang [6], Chazarain [3] and De Paris [5]. Concerning the above things, an article of the author will give the proof. Observe that, this is fairly evident from Theorem 5 in Lecture III. Levi condition is also called Levi-Lax condition.

Now let us return to the case of Gevrey class. According to Theorem 2, assuming all the coefficients belong to $\gamma^{(s)}$, we have

Theorem 3. In order that the Cauchy problem in Gevrey class $\gamma^{(s)}$ be uniformly wellposed, it it necessary that s satisfies

$$\frac{1}{s} \geq \max(\rho_1, \rho_2, \cdots, \rho_k) = \rho.$$

Note. Komatsu defined the notion of irregularity [14]. This number is closely connected with ours. Theorem 3 is not proved here. A forthcoming paper will give its proof [29]. Observe that the corresponding result is given by Ivrii [27].

What about the sufficiency ? We can prove the following

Theorem 4. If $\frac{1}{s} > \rho$, then there exists a unique solution $u(x,t)$ for $t \in [t_0, T]$. If $\frac{1}{s} = \rho$, then there exists a local solution (in t).

Note. Theorem 4 is not entirely new. The original result of Ohya [22], and recent result of Komatsu [14] give already essentially the same result. However, the author wishes to insist that the number ρ is an essential index by Theorem 3 and Theorem 4.Moreover the proof is different from theirs

§ 2. Fundamental proposition.

To simplify the arguments, we consider first

$$\begin{cases} \left(\partial_t - i\lambda(x,t;D) - a(x,t;D)\right) u = f, & (x,t) \in \mathbb{R}^n \times [0, T] \\ u\Big|_{t=0} = 0, & \end{cases} \tag{9}$$

Let us remark here that we denoted the dimension of the space by n. The assumption is the following.

i) $\lambda(x,t;\xi)$ is real-valued, of homogeneous degree 1 in ξ.

ii) $a(x,t;\xi) \in S^{\rho}_{1,0}$, $0 < \rho < 1$.

iii) λ and a satisfy the following Gevrey estimates

$$|a^{(\alpha)}_{(\beta)}| \leq C_A \alpha! \beta!^s \, C_0^{|\beta|} (|\xi| + 1)^{\rho - |\alpha|} \quad \text{for } |\alpha| \leq 2n,$$

$$|\lambda^{(\alpha)}_{(\beta)}| \leq C_\lambda \alpha! \beta!^s \, C_0^{|\beta|} |\xi|^{1-|\alpha|} \quad \text{for } |\alpha| \leq 2n .$$

iv) $a(x,t;\xi)$, $\lambda(x,t;\xi)$ are continuous in t with all its derivatives in (x, ξ).

v) We assume $f(\cdot,t) \in \mathcal{E}^0_t([0, T]; \gamma^{(s)}_{L^2})$, where $f \in \gamma^{(s)}_{L^2}$ means

$$\| \partial^\alpha f(x) \|_{L^2} \leq c_0 \alpha!^s \, C^{|\alpha|}, \quad \alpha \geq 0 \quad ({}^\exists c_0, C > 0).$$

We prepare several lemmas.

Lemma 1. Let $0 \leq \rho \leq 1$, and $\| f \|_\rho = \left(\int |\xi|^{2\rho} |\hat{f}(\xi)|^2 \, d\xi \right)^{\frac{1}{2}}$. Then

$$\| f \|_\rho \leq \| f \|_0^{1-\rho} \| f \|_1^{\rho} .$$

Lemma 2. $\forall a \geq 0,$

$$a!^{1-\rho}(a+1)!^{\rho} \leq c(\rho)\cdot(a+\rho)!$$

where $c(\rho)$ is independent of a.

Lemma 3. Let $a(x,\xi) \in S^{\rho}_{0,0}$. More precisely

$$|a^{(\alpha)}_{(\beta)}| \leq C_a|\alpha|!^s\, C_0^{|\alpha|}(|\xi|+1)^{\rho}, \quad \forall \alpha \geq 0, \quad |\beta| \leq n+2$$

and $\sum_{|\alpha|=m} \|\partial^{\alpha}v\| \leq M(k+m)!^s\, C^m, \quad \forall m \geq 0,$ k being a non-negative number (not necessarily integer), then there exists a positive constant K_a (depending only on a) such that

$$\sum_{|\alpha|=m} \|\partial^{\alpha}(av)\| \leq K_a M\,(k+m+\rho)!^s\, C^{m+\rho}, \quad \forall m \geq 0,$$

provided that $C \geq 4C_0, \quad C \geq 1.$

Preparing these lemmas, we consider the mapping $f \longmapsto u$ in (9), in the space $\gamma^{(s)}$. We solve (9) by successive approximations:

$$u = u_0 + u_1 + \cdots + u_j + \cdots$$

where

$$\begin{cases} (\partial_t - i\lambda)\,u_0 = f, & \text{with } u_0\big|_{t=0} = 0 \\ (\partial_t - i\lambda)\,u_1 = au_0, & \text{with } u_1\big|_{t=0} = 0 \\ \quad\cdots & \quad\cdots \\ (\partial_t - i\lambda)\,u_j = au_{j-1}, & \text{with } u_j\big|_{t=0} = 0 \end{cases} \tag{10}$$

For our purpose, we consider first the estimates of successive derivatives of $\| \partial^{\alpha} u \|$ of the equation

$$\left\{ \begin{aligned} & \left(\partial_t - i\lambda(x,t;D)\right) u = f \\ & u\Big|_{t=0} = 0 \, . \end{aligned} \right. \tag{11}$$

To make precise the estimate, let us explain this fairly in detail. The proof is given in the Appendix. First decompose

$$\lambda(x,t;D) = \lambda_0(x,t;D)\Lambda \quad (\hat{\Lambda} = |\xi|),$$

where we have

$$\| \partial_x^{\alpha} \lambda_0(x,t;D) \|_{\mathcal{L}(L^2;L^2)} \le C(\lambda_0) \cdot (|\alpha| - 1)!^s \, C_0^{|\alpha|-1} \tag{12}$$

for $|\alpha| \ge 1$. (See (A.9) in Lecture III .)

Next, we have from (11),

$$\frac{d}{dt} \| u(\cdot,t) \| \le \gamma_0 \| u(\cdot,t) \| + \| f(\cdot,t) \| , \tag{13}$$

γ_0 being determined by λ_0. Define

$$\begin{aligned} & K_0(t) = e^{\gamma_0 t}, \quad K(t) = (1 + \gamma t)\, e^{\gamma t}, \\ & \text{where } \gamma = C(\lambda_0) \max\left(n^2, 2^n n\right). \end{aligned} \tag{14}$$

Put

$$f_m(t) = \sum_{|\alpha|=m} \| \partial_x^{\alpha} f(\cdot,t) \| , \quad u_m(t) = \sum_{|\alpha|=m} \| \partial_x^{\alpha} u(\cdot,t) \| .$$

Then we have

Lemma 4. If

$$f_m(t) \le A_f\ K_0(t)\ K(t)^{m+k}\ C^{m+k}\ (m+k)!^s\ \frac{t^j}{j!} \tag{15}$$

($m = 0, 1, 2, \cdots$), where $k \ge 0$ is any real number, and $j \ge 0$ is an integer. Then, if $C \ge 4C_0$ (Note that C_0 appears in (12)), we have

$$u_m(t) \le 2A_f\ K_0(t)\ K(t)^{m+k}\ C^{m+k}\ (m+k)!^s\ \frac{t^{j+1}}{(j+1)!}. \tag{16}$$

Lemma 4 is crucial in our arguments. Its proof is given in the Appendix.

Now we return to (10). Without loss of generality, we can assume (15) for f, with $A_f = 1$, $k = j = 0$, and $C \ge 4C_0$ (see (16)). Then, by Lemma 4,

$$\sum_{|\alpha|=m} \| \partial^\alpha u_0(\cdot,t) \| \le 2K_0(t)\ \{C\,K(t)\}^m\ m!^s\ t. \tag{17}$$

Next, by Lemma 3,

$$\sum_{|\alpha|=m} \| \partial^\alpha (a u_0) \| \le 2K_a\ K_0(t)\ \{CK(t)\}^{m+\rho}\ (m+\rho)!^s\ t.$$

Thus, again by Lemma 4,

$$\sum_{|\alpha|=m} \| \partial^\alpha u_1 \| \le 2^2\ K_a\ K_0(t)\ \{CK(t)\}^{m+\rho}\ (m+\rho)!^s\ \frac{t^2}{2!}. \tag{18}$$

The same arguments can be applied successively to $u_2, u_3, \cdots$.
For reader's convenience, we repeat it for u_2. First by Lemma 3,

$$\sum_{|\alpha|=m} \| \partial^\alpha (au_1) \| \le 2^2 K_a^2 n^2 K_0(t) \{CK(t)\}^{m+2\rho} (m + 2\rho)!^s \frac{t^2}{2!}.$$

Thus by Lemma 4,

$$\sum_{|\alpha|=m} \| \partial^\alpha u_2 \| \le 2^3 K_a^2 K_0(t) \{CK(t)\}^{m+2\rho} (m + 2\rho)!^s \frac{t^3}{3!}. \tag{19}$$

We see that

$$\sum_{|\alpha|=m} \| \partial^\alpha u_j \| \le 2^{j+1} K_a^j K_0(t) \{CK(t)\}^{m+j\rho} (m + j\rho)!^s \frac{t^{j+1}}{(j + 1)!}. \tag{20}$$

Thus

$$\sum_{j=0}^\infty \sum_{|\alpha|=m} \| \partial^\alpha u_j \| \le 2K_0(t) \{CK(t)\}^m t \sum_{j=0}^\infty (2K_a)^j \{CK(t)\}^{j\rho} (m + j\rho)!^s \frac{t^j}{(j + 1)!}.$$

Now, there exists (an absolute) constant C_0' such that

$$(m + j\rho)!^s \le C_0'^{\,m+j\rho}\, m!^s\, (j\rho)!^s.$$

Next, there exists (an absolute) constant C' such that

$$(j\rho)!^s \le C'^j\, j!^{\rho s},$$

and substituting $(j + 1)!$ by $j!$, we see that the sum $\sum_{j=0}^\infty \cdots$ is estimated by

$$C_0'^m\, m!^s \sum_{j=0}^{\infty} \left(2C'K_a\right)^j \left\{CC_0'K(t)\right\}^{\rho j} \frac{t^j}{j!^{1-\rho s}} .$$

We denote

$$\Phi(t;C) = 2 \sum_{j=0}^{\infty} \left(2C'K_a\right)^j \left\{CC_0'K(t)\right\}^{\rho j} \frac{t^j}{j!^{1-\rho s}} .$$

Let us recall that we assumed in (15) $k = 0$ and $j = 0$. In general case, replacing here m by $m + k$, we see easily that the following estimate holds:

$$\sum_{j=0}^{\infty} \sum_{|\alpha|=m} \| \partial^{\alpha} u_j \| \le A_f \left\{\Phi(t;C) C_0'^{\,m+k}\right\} K_0(t) \left\{CK(t)\right\}^{m+k} (m+k)!^s \frac{t^{j+1}}{(j+1)!} . \tag{21}$$

Summing up, we have obtained

Fundamental Proposition.

The Cauchy problem for (9), with $f(\cdot,t) \in \mathcal{E}_t^0([0, T];\gamma^{(s)}_{L^2})$ has a unique solution $u(\cdot,t) \in \mathcal{E}_t^1([0, T];\gamma^{(s)}_{L^2})$ provided that $1 - \rho s > 0$. More precisely, if we assume for f the same estimate as (15), then we have

$$u_m(t) \le A_f \left\{\Phi(T;C)\, C_0'^{\,m+k}\right\} K_0(t) \left\{CK(t)\right\}^{m+k} (m + k)!^s \frac{t^{j+1}}{(j + 1)!} \tag{22}$$

$(m = 0, 1, 2, \cdots)$.

In the case $1 - \rho s = 0$, $\Phi(t;C)$ has positive radius of convergence t_0'. Thus for $t < t_0$, the solution $u(\cdot,t)$ exists. It has the same estimate as above, replacing $\Phi(T;C)$ by $\Phi(t;C)$.

Note. The uniqueness follows from the consideration of P^*. P^* has the same property as P. Thus, under the restriction $u(\cdot,t) \in \mathcal{E}_t^1([0, T];L^2)$, the uniqueness holds. Finally, observe that if the initial data at $t = 0$, $u_0(x) \in \gamma^{(s)}_{L^2}$ is given, then by taking $v(\cdot,t) = u(\cdot,t) - u_0(x)$, as a new unknown function, we have the existence theorem.

§ 3. Proof of Theorem 4.

Now we pass to Theorem 4. For simplicity, we assume for P, $k = 2$ (see (4) and (6)). More precisely,

$$P = P_2 \circ P_1 + R \tag{23}$$

$$\begin{cases} P\,u = f \\ \partial_t^{\,j} u\big|_{t=0} = 0 \quad (0 \le j \le m-1)\,. \end{cases} \tag{24}$$

Observe that, for the perfect factorization (in $\gamma^{(s)}$), we need to assume the coefficients of P are of $\gamma^{(s)}$ with respect to (x,t), and moreover the perfect factorization in Gevrey class is delicate. Here we assume the coefficients of P are merely of C^∞ in t and, of course, of Gevrey class $\gamma^{(s)}$ in x, or more precisely (as for t) sufficiently differentiable (however, we do not make precise up to what order of differentiability in t). Instead of the perfect factorization, we use the following easy factorization. A factorization of the form (6) (or (3)) is a <u>good factorization</u> in $\gamma^{(s)}$, if R satisfies merely the following condition: Denoting $R(x,t;D) = \sum_{j=0}^{m-1} r_j(x,t;D)\,\partial_t^{\,j}$, each $r_j(x,t;D)$ satisfies

$$|\partial_x^{\alpha}\partial_\xi^{\beta} r_j(x,t;\xi)| \le \tilde{C}\cdot\alpha!^s\, C_0^{|\alpha|}(1 + |\xi|)^{-m-|\beta|}, \quad \forall \alpha \ge 0, \quad |\beta| \le n+2 \tag{25}$$

Let us remark that order $r_j \leq -m$. This factorization is carried out in the same way as in C^∞-case (see [15], [24]).

In order to reduce the problem (24) to the Fundamental Proposition, it is convenient to consider the following equivalent system. Put

$$\left\{\begin{aligned} u_0 &= u \\ u_1 &= (\partial_t - i\lambda_1)\, u \\ &\;\vdots \\ u_{m_1-1} &= (\partial_t - i\lambda_1)^{m_1-1}\, u \end{aligned}\right. \qquad \left\{\begin{aligned} u_{m_1} &= P_1\, u \\ u_{m_1+1} &= (\partial_t - i\lambda_2)\, u_{m_1} \\ u_{m_1+2} &= (\partial_t - i\lambda_2)^2\, u_{m_1} \\ &\;\vdots \\ u_{m-1} &= (\partial_t - i\lambda_2)^{m_2-1}\, u_{m_1}. \end{aligned}\right. \tag{26}$$

Observe that $P_1\, u$ can be written as (see (7)),

$$P_1\, u = (\partial_t - i\lambda_1)u_{m_1-1} + a_{1,1}\, u_{m_1-1} + a_{2,1}\, u_{m_1-2} + \cdots + a_{m_1,1}\, u_0.$$

In the same way,

$$P_2 \circ P_1\, u = (\partial_t - i\lambda_2)u_{m-1} + a_{1,2}\, u_{m-1} + a_{2,2}\, u_{m-2} + \cdots + a_{m_2,2}\, u_{m_1}.$$

Thus the equation (24) is equivalent to

$$\left\{\begin{aligned} &(\partial_t - i\lambda_1)u_j = u_{j+1} \qquad (0 \leq j \leq m_1 - 2) \\ &(\partial_t - i\lambda_1)u_{m_1-1} = -a_{m_1,1}\, u_0 - a_{m_1-1,1}\, u_1 - \cdots - a_{1,1}\, u_{m_1-1} + u_{m_1} \\ &(\partial_t - i\lambda_2)u_j = u_{j+1} \qquad (m_1 \leq j \leq m - 2) \\ &(\partial_t - i\lambda_2)u_{m-1} = -a_{m_2,2}\, u_{m_1} - a_{m_2-1,2}\, u_{m_1+1} - \cdots - a_{1,2}\, u_{m-1} - Ru + f. \end{aligned}\right. \tag{27}$$

In the last relation, Ru appears. We want to see the relation between $(u_0, u_1, \dots, u_{m-1})$ and $(u, \partial_t u, \dots, \partial_t^{m-1} u)$. We have

<u>Lemma 5</u>.

$$\begin{gathered}\partial_t^{\,j} u = C^j_j\, u_0 + C^j_{j-1}\, u_1 + \cdots + C^j_1\, u_{j-1} + u_j, \quad (0 \le j \le m-1), \\ \text{order } C^j_k(x,t;\xi) \le k, \quad C^j_k(x,t;\xi) \in S^k_{1,0} \cap \gamma^{(s)}_x .\end{gathered} \tag{28}$$

Conversely,

$$\begin{gathered}u_j = d^j_j\, u + d^j_{j-1}\, \partial_t u + \cdots + d^j_1\, \partial_t^{\,j-1} u + \partial_t^{\,j} u, \quad (0 \le j \le m-1) \\ \text{order } d^j_k(x,t;\xi) \le k, \quad d^j_k(x,t;\xi) \in S^k_{1,0} \cap \gamma^{(s)}_x .\end{gathered} \tag{29}$$

<u>Proof</u>. First observe that, for general k, $(\partial_t - i\lambda_1)^k u$ is expressed in the form,

$$(\partial_t - i\lambda_1)^k u = \partial_t^{\,k} u + g^k_1(x,t;D)\partial_t^{\,k-1} u + g^k_2(x,t;D)\partial_t^{\,k-2} u + \cdots + g^k_k(x,t;D)u$$

where g^k_j is of order $\le j$, and its symbol of Gevrey class s with respect to x. For $1 \le k \le m_1 - 1$, we see (28) by induction on k. In fact, the above formula and the hypothesis on induction show that

$$\begin{aligned}\partial_t^{\,k} u = u_k &- g^k_1\left(c^{k-1}_{k-1}\, u_0 + c^{k-1}_{k-2}\, u_1 + \cdots + c^{k-1}_1\, u_{k-2} + u_{k-1}\right) \\ &- g^k_2\left(c^{k-2}_{k-2}\, u_0 + c^{k-2}_{k-3}\, u_1 + \cdots + c^{k-2}_1\, u_{k-3} + u_{k-2}\right) \\ &\quad \cdots \\ &- g^k_k\, u_0 .\end{aligned}$$

Next, by definition,

$$u_{m_1} = (\partial_t - i\lambda_1)^{m_1} u + \sum_{k=1}^{m_1} a_{k,1}(\partial_t - i\lambda_1)^{m_1-k} u .$$

Namely,

$$(\partial_t - i\lambda_1)^{m_1} u = u_{m_1} - \sum_{k=1}^{m_1} a_{k,1}\, u_{m_1-k} .$$

Hence

$$\partial_t^{m_1} u = u_{m_1} - \sum_{k=1}^{m_1} a_{k,1}\, u_{m_1-k} - \left(g_1^{m_1}\partial_t^{m_1-1} u + g_2^{m_1}\partial_t^{m_1-2} u + \cdots + g_{m_1}^{m_1} u\right).$$

Since order $a_{k,1} \le k\rho$ ($< k$), we verify (28).

Next,

$$u_{m_1+1} = (\partial_t - i\lambda_2)u_{m_1} = (\partial_t - i\lambda_2)(\partial_t - i\lambda_1)^{m_1} u + \sum_{k=1}^{m_1} (\partial_t - i\lambda_2)a_{k,1}(\partial_t - i\lambda_1)^{m_1-k} u.$$

From this relation, we can verify (28) for $\partial_t^{m_1+1} u$. For $m_1 + 2 \le k \le m - 1$, the same arguments can be applied. Finally (29) is derived from (28) easily.

Now by Lemma 5, we can express Ru in the form:

$$\begin{aligned} Ru &= \sum_{j=0}^{m-1} r_j(x,t;D)\left(c_j^j\, u_0 + c_{j-1}^j\, u_1 + \cdots + c_1^j\, u_{j-1} + u_j\right) \\ &= \sum_{j=0}^{m-1} b_j(x,t;D)\, u_j, \end{aligned} \tag{30}$$

where, in view of (25),

$$b_j(x,t;\xi) \in S_{1,0}^{-1} \cap \gamma_x^{(s)} .$$

<u>Note</u>. The differentiability in t of the coefficients of P are reflected in that of $a_{k,j}(x,t;\xi)$ and that of $b_j(x,t;\xi)$. If we assume the coefficients are C^∞ in t, then this is the same with $a_{k,j}$ and b_j.

Now denoting $U = {}^t(u_0, u_1, \cdots, u_{m-1})$, the equation (24) becomes

$$\partial_t U = \left(\begin{array}{c|c} D_1 & \bigcirc \\ & 1 \\ \hline \bigcirc & D_2 \end{array}\right) U - BU + F \tag{31}$$

where

$$D_j = \begin{pmatrix} i\lambda_j & 1 & & & \\ & i\lambda_j & 1 & & \bigcirc \\ & & \ddots & \ddots & \\ & \bigcirc & & \ddots & 1 \\ -a_{m_j,j} & -a_{m_j-1,j} & \cdots & & i\lambda_j - a_{1,j} \end{pmatrix} \quad (j=1,2), \quad B = \begin{pmatrix} & & \bigcirc & \\ & & & \\ b_0 & b_1 & \cdots & b_{m-1} \end{pmatrix},$$

and $F = {}^t(0, 0, \cdots, 0, f)$. Observe that D_j is $m_j \times m_j$ matrix, and order $a_{k,j} \leq k\rho$.

We consider (31) with a weight. Put

$$\tilde{u}_j(\cdot,t) = \langle\Lambda\rangle^{(m-1-j)\rho}\, u_j(\cdot,t) \qquad (0 \leq j \leq m-1) \tag{32}$$

where $\widehat{\langle\Lambda\rangle u}(\xi) = (1 + |\xi|^2)^{\frac{1}{2}}\, \tilde{u}(\xi)$. Let us see, to what system, (31) is transformed.

$$\partial_t u_j = i\lambda u_j + u_{j+1} \qquad (j \neq m_1 - 1, m - 1),$$

λ being one of λ_1, λ_2. This becomes

$$\partial_t \tilde{u}_j = i\lambda\tilde{u}_j + \langle\Lambda\rangle^{\rho}\tilde{u}_{j+1} + [\langle\Lambda\rangle^{(m-1-j)\rho}, i\lambda]\langle\Lambda\rangle^{-(m-1-j)\rho}\,\tilde{u}_j,$$

where the last term is of order 0. Next consider the m_1-th row

$$\langle\Lambda\rangle^{m_2\rho} a_{m_1-j,1}\, u_j = \langle\Lambda\rangle^{m_2\rho} a_{m_1-j,1}\, \langle\Lambda\rangle^{-(m-1-j)\rho}\,\tilde{u}_j.$$

We see that the order of the right hand pseudo-differential operator is of order $\leq \rho$. It is the same with the last row. Indeed (see (27)),

$$a_{m_2-j,2}\, u_{m_1+j} = a_{m_2-j,2}\, \langle\Lambda\rangle^{(m_2-1-j)\rho}\,\tilde{u}_{m_1+j},$$

and the order of the operator of the right-hand side is $\leq \rho$. Observe that B is transformed to $\tilde{B}$ whose order is ≤ -1.

Thus we are led to the following system, denoting $\tilde{U} = {}^t(\tilde{u}_0, \tilde{u}_1, \cdots, \tilde{u}_{m-1})$,

$$\partial_t\tilde{U} = i\begin{pmatrix} \lambda_1 & & & & & \\ & \lambda_1 & & & \bigcirc & \\ & & \ddots \; \lambda_1 & & & \\ & & & \lambda_2 & & \\ & \bigcirc & & & \ddots & \\ & & & & & \lambda_2 \end{pmatrix}\tilde{U} + A(x,t;D)\,\tilde{U} + F \tag{32}$$

where $A = \left(a_{ij}(x,t;D)\right)_{1\leq i,j\leq m}$, and

$$a_{ij}(x,t;\xi) \in S^{\rho}_{1,0} \cap \gamma_x^{(s)}. \tag{33}$$

Now we see that (32) is a straightforward extension of the Cauchy problem (9). More precisely, the arguments used to prove Fundamental Proposition can be applied to (32) without any significant modification. The proof is left to the reader. What about the proof of the uniqueness ? The uniqueness holds under the assumption $u(\cdot,t) \in \mathcal{E}^m([0, T];L^2)$, or even $u(\cdot,t) \in \mathcal{E}_t^m([0, T];(H^\infty)')$ ($(H^\infty)'$ is the dual space of H^∞). This is carried out by considering the adjoint system to (32), because the adjoint system satisfies the same assumption as (32).

§ 4. Gevrey property in t of solutions.

We want to show the following fact: When the coefficients belong to $\gamma^{(s)}_{x,t}$, namely we assume all the coefficients belong to $\gamma^{(s)}$ as function in (x,t) and also we assume the same property for $f(x,t)$, then the solution $u(x,t)$, which is announced in Theorem 4, belongs also to $\gamma^{(s)}_{x,t}$. As we mentioned at the end of § 3, it will be enough to show the property for (9):

$$\partial_t u(x,t) = \big(i\lambda(x,t;D) + a(x,t;D)\big)\, u(x,t) + f.$$

Putting $i\lambda + a = a_1$, rewrite this in the form:

$$\partial_t\, u = a_1\, u + f. \tag{34}$$

First from the assumption on λ and a, we can assume for a_1 the following estimates: $\exists\, A,\ C_0\ (>0)$ such that, denoting $a_1\langle\Lambda\rangle^{-1} = a_0(x,t;D)$, we have

$$\left\| \partial_x^{\alpha} \partial_t^{k} a_0(x,t;D) \right\|_{\mathcal{L}(L^2;L^2)} \leq A(k + |\alpha|)!^s \, C_0^{k+|\alpha|}. \tag{35}$$

We assume also

$$\| \partial_x^{\alpha} \partial_t^{k} f(\cdot,t) \| \leq M(k + |\alpha|)!^s \, C_1^{k+|\alpha|} \qquad (2 \leq 2C_0 \leq C_1). \tag{36}$$

We can assume moreover, in view of Fundamental Proposition (see (22)),

$$\| \partial_x^{\alpha} u(\cdot,t) \| \leq M' |\alpha|!^s \, C_2^{|\alpha|} \qquad (\exists M' \geq 2M). \tag{37}$$

Under these assumptions, we want to show the following estimates: For all $\alpha \geq 0$, and $k \geq 0$,

$$\| \partial_x^{\alpha} \partial_t^{k} u(\cdot,t) \| \leq M'(k + |\alpha|)!^s \, C_2^{|\alpha|} \, C_3^{k}, \tag{38}$$

taking C_3 sufficiently large.

The basic idea of the proof is simple. We differentiate (34) k times in t,

$$\partial_t^{k+1} u = \sum_{0 \leq k' \leq k} \binom{k}{k'} \partial_t^{k'} a_1 \cdot \partial_t^{k''} u + \partial_t^{k} f \, . \tag{39}$$

Then assuming (38) for $0 \leq k'' \leq k$, we show (38) for $k = k + 1$. It would be convenient to use the following.

Lemma 6. Let

$$\| \partial_x^{\alpha} b(x;D) \| \le A(k + |\alpha|)!^s \, C_0^{|\alpha|}, \quad \alpha \ge 0,$$

$$\| \partial^{\alpha} v \| \le B(k' + |\alpha|)!^s \, C^{|\alpha|} \qquad (C \ge C_0),$$

k, k' being non-negative integers. Then

$$\| \partial^{\alpha}(bv) \| \le AB\, c(s - 1)\cdot(k + k' + |\alpha|)!^s \, C^{|\alpha|} \Big/ \binom{k + k'}{k}^s ,$$

where, for $\varepsilon > 0$, $c(\varepsilon)$ is defined by

$$c(\varepsilon) = \sup_m \sum_{m'=0}^{m} \binom{m}{m'}^{-\varepsilon}. \tag{40}$$

Note. $1 < c(\varepsilon) < +\infty$.

The proof of this lemma is given in Appendix.

First,

$$\partial_t^{k'} a_1 (\partial_t^{k''} u) = \partial_t^{k'} a_1 \Lambda^{-1} (\langle\Lambda\rangle \partial_t^{k''} u) = \partial_t^{k'} a_0(x,t;D) (\langle\Lambda\rangle \partial_t^{k''} u).$$

Using (38),

$$\left\| \partial_x^{\alpha} \langle\Lambda\rangle \partial_t^{k''} u \right\|^2 = \left\| \partial_x^{\alpha} \partial_t^{k''} u \right\|^2 + \sum_{j=1}^{n} \left\| \partial_x^{\alpha+\ell_j} \partial_t^{k''} u \right\|^2 .$$

$$\begin{aligned}
\left\| \partial_x^{\alpha} \langle\Lambda\rangle \partial_t^{k''} u \right\| &\le \left\| \partial_x^{\alpha} \partial_t^{k''} u \right\| + \sum_j \left\| \partial_x^{\alpha+\ell_j} \partial_t^{k''} u \right\| \\
&\le M'(k''+|\alpha|)!^s C_2^{|\alpha|} C_3^{k''} + \sum_j M'(k''+1+|\alpha|)!^s C_2^{|\alpha|+1} C_3^{k''} \\
&\le M'(n+1)(k''+1+|\alpha|)!^s \, C_2^{|\alpha|+1} \, C_3^{k''}.
\end{aligned}$$

If we take account of (35), Lemma 6 yields

$$\left\| \partial_x^{\alpha}(\partial_t^{k'} a_1 \cdot \partial_t^{k''} u) \right\| \leq AM'(n+1)c(s-1)\cdot(k+1+|\alpha|)!^s\, C_2^{|\alpha|+1}\, C_3^{k} \Big/ \binom{k+1}{k'}^s .$$

Thus by (39),

$$\| \partial_x^{\alpha} \partial_t^{k+1} u \| \leq \sum_{k'=0}^{k} \binom{k}{k'} \| \cdots \| + \| \partial_x^{\alpha} \partial_t^{k} f \|$$

$$\leq AM'(n+1)c(s-1)^2 \cdot (k+1+|\alpha|)!^s C_2^{|\alpha|+1} C_3^{k} + M(k+|\alpha|)!^s C_1^{|\alpha|+k}$$

$$= M'(k+1+|\alpha|)!^s C_2^{|\alpha|} C_3^{k+1} \left[A(n+1)c(s-1)^2 C_2/C_3 + \frac{M}{M'} \frac{1}{(k+1+|\alpha|)^s} \left(\frac{C_1}{C_2}\right)^{|\alpha|} \cdot \frac{C_1^{k}}{C_3^{k+1}} \right] .$$

Thus, if we take

$$C_3 = 2A\,(n+1)\,c(s-1)^2\, C_2,$$

then taking account of $M/M' \leq 1/2$, $[\cdots] \leq 1$, we can prove (38). In the case $k = 1$, (38) is verified. In fact, using (37)

$$\| \partial_x^{\alpha} \partial_t u \| \leq AM'(n+1)c(s-1)\cdot(|\alpha|+1)!^s\, C_2^{|\alpha|+1} + M|\alpha|!^s\, C_1^{|\alpha|}.$$

This estimate coincides with the former one with $k = 0$ if we replace there $c(s-1)^2$ by $c(s-1)$. Since $c(s-1) > 1$, (38) holds when $k = 1$.

§ 5. Comments.

In 1975, Ivrii discussed the Cauchy problem in Gevrey class removing the assumption of constant multiplicity [10]. Next, Bronstein succeeded in removing the assumption completely [2]. Concerning these, see more recent works of Kajitani [11], [12], and Nishitani [21].

Although we are mainly concerned with the Cauchy problem in Gevrey class, we explain briefly the problem in C^∞-class. First, the sufficient part of Theorem 2 is almost evident in view of the proof of Theorem 4. In fact, in (31), the condition order $a_{k,j}(x,t;\xi) \leq 0$ implies that we can regard this system already in the form (32) ($\tilde{U} = U$), with $a_{ij}(x,t;\xi) \in S^0_{1,0}$ ($1 \leq i,j \leq m$). The same arguments (but much simpler) as in the Gevrey class can be applied taking H^m as a basic space (m is an arbitrary positive number).

Next, in the case of non-strictly hyperbolic systems, there are several works: Vaillant [25], Kajitani [13], Demay [4], Yamahara [26] etc.. They are treating the problem from different points of view. Contrary to the case of single equations, the problem is not simple even if we suppose the constancy of multiplicity. One could say that this problem is still open. The work of Matsumoto [18] reveals its complex nature.

Appendix

§ A.1 Proof of Lemma 4.

In considerations of Gevrey class, it seems to the author, one can not avoid entirely tedious and sometimes delicate computation. The following proof will be one of such examples.

We differentiate (11) with respect to x to obtain

$$(\partial_t - i\lambda)(\partial^\alpha u) = \partial^\alpha f + i \sum_{\substack{\alpha' \le \alpha \\ \alpha' > 0}} \binom{\alpha}{\alpha'} \partial^{\alpha'}\lambda \cdot \partial^{\alpha''} u, \qquad (\alpha' + \alpha'' = \alpha). \tag{A.1}$$

Using (12),

$$\begin{aligned} \| \partial^{\alpha'}\lambda \cdot \partial^{\alpha''} u \| &\le \| \partial^{\alpha'}\lambda_0 \|_{\mathcal{L}(L^2;L^2)} \| \Lambda \partial^{\alpha''} u \| \\ &\le C(\lambda_0)\left(|\alpha'| - 1\right)!^s \, C_0^{|\alpha'|-1} \sum_{j=1}^{n} \| \partial^{\alpha''+e_j} u \| . \end{aligned} \tag{A.2}$$

Thus, from (13),

$$\frac{d}{dt} \| \partial^\alpha u \| \le \gamma_0 \| \partial^\alpha u \| + \| \partial^\alpha f \| + \sum_{\alpha'} \binom{\alpha}{\alpha'} \| \partial^{\alpha'} \lambda_0 \| \sum_{j=1}^{n} \| \partial^{\alpha''+e_j} u \| .$$

Summing up this inequality for all $|\alpha| = m$,

$$\frac{d}{dt} u_m(t) \le \gamma_0 u_m(t) + f_m(t) + \sum_{|\alpha|=m} \sum_{\alpha'} \binom{\alpha}{\alpha'} \| \partial^{\alpha'} \lambda_0 \| \sum_{j=1}^{n} \| \partial^{\alpha''+e_j} u \| . \tag{A.3}$$

Now we consider the estimate of the coefficients of $\| \partial^\beta u \|$ which appear in the last sum.

i) Case $|\beta| = m$. β is fixed. Then, the number of α'', such that $\alpha'' + e_j = \beta$ is at most n. Next, when such an α'' is fixed, then the number of α ($|\alpha| = m$) such that $\alpha > \alpha''$ is at most n. Since $\binom{\alpha}{\alpha'} \le m$, the coefficient of $\| \partial^\beta u\|$ in (A.3) is estimated by $c(\lambda_0)mn^2$.

ii) Case $|\beta| = m - p$, $p \ge 1$. We fix such a β. Since $\beta - e_j = \alpha''$, the number of α such that $\beta - e_j$ ($= \alpha''$) $< \alpha$, namely that of $\alpha - (\beta - e_j) > 0$ is equal to the number of $(\nu_1, \cdots, \nu_n)$ $(\nu_j \ge 0)$ such that $\nu_1 + \cdots + \nu_n = p + 1$. This number is equal to $C_{n-1}^{p+1+n-1}$, which is estimated by 2^{p+n}. Now $\binom{\alpha}{\alpha'} \le \binom{|\alpha|}{|\alpha'|} = \binom{m}{p+1}$.

Thus, in (A.3), the coefficient of $\sum_{|\beta|=m-p} \| \partial^\beta u\|$ $(1 \le p \le m - 1)$ is estimated by

$$\binom{m}{p+1} c(\lambda_0)\ p!^s\ C_0^p\ n\ 2^{p+n}.$$

Summing up, (A.3) becomes

$$\begin{aligned} \frac{d}{dt} u_m(t) \le \gamma_0\ u_m(t) + c(\lambda_0)\ n^2 m\ u_m(t) + f_m(t) \\ + \sum_{p=1}^{m-1} \binom{m}{p+1} c(\lambda_0)\ p!^s\ C_0^p\ n\ 2^{p+n}\ u_{m-p}(t). \end{aligned} \tag{A.4}$$

Now, we are ready to prove Lemma 4 by induction on m. In the case $m = 0$, this is clear, because

$$\frac{d}{dt} u_0(t) \le \gamma_0\ u_0(t) + f_0(t) \quad \text{implies}$$

$$u_0(t) \le \int_0^t \exp[\gamma_0(t-s)]\ f_0(s)\ ds \le A_f \int_0^t \exp[\gamma_0(t-s)]\ K_0(s)\left(K(s)C\right)^k \frac{s^j}{j!}\ ds$$

$$\le A_f\ K_0(t)\left(K(t)C\right)^k \frac{t^{j+1}}{(j+1)!}.$$

In the case $m = 1$, the proof is immediate, because we have

$$\frac{d}{dt} u_1(t) \le \left(\gamma_0 + c(\lambda_0)\ n^2\right) u_1(t) + f_1(t),$$

and since $\gamma \ge c(\lambda_0)\ n^2$ (see (14)), we get

$$u_1(t) \le A_f \int_0^t \exp[\gamma_0(t-s)]\ \exp[\gamma(t-s)]\ K_0(s)\left(K(s)C\right)^{k+1} \frac{s^j}{j!}\ ds$$

$$\le A_f\ K_0(t)\left(K(t)C\right)^{k+1} \frac{t^{j+1}}{(j+1)!}\ .$$

Now, we pass to general m. By the assumption on induction on m, we have

$$u_{m-p}(t) \le 2A_f\ K_0(t)\left(K(t)C\right)^{k+m-p}(k+m-p)!^s\ \frac{t^{j+1}}{(j+1)!} \qquad (1 \le p \le m-1). \tag{A.5}$$

Let us show that (A.5) implies that the last sum in (A.4) is estimated by

$$A_f\left(c(\lambda_0)n2^n\right)2C_0(m-1)(k+m)!^s\ K_0(t)\left(K(t)C\right)^{k+m-1} \frac{t^{j+1}}{(j+1)!}\ . \tag{A.6}$$

First,

$$\binom{m}{p+1} p!^s\, C_0^p\, 2^p \left(K(t)C\right)^{-p+1} (k+m-p)!^s = 2C_0 \binom{m}{p+1} p!^s (k+m-p)!^s \left(\frac{2C_0}{K(t)C}\right)^{p-1}.$$

$$\binom{m}{p+1} = \frac{m-p}{p+1}\binom{m}{p} \le \frac{m-1}{2}\binom{m}{p} \le \frac{m-1}{2}\binom{k+m}{p}.$$

Thus, $\binom{m}{p+1} p!^s (k+m-p)!^s \le \frac{m-1}{2}\binom{k+m}{p} p!^s (k+m-p)!^s$. Since $p!(k+m-p)!\binom{k+m}{p} = (k+m)!$, the last is estimated by $(\frac{m-1}{2})\cdot(k+m)!^s$. Next, taking account of $2C_0 / K(t)C \le \frac{1}{2}$, we have $\sum_{p=1}^{\infty}\left(2C_0 / K(t)C\right)^{p-1} \le 2$. Thus (A.6) is verified.

Now, since $c(\lambda_0)\, n\, 2^n \le \gamma$ in (A.6), (A.4) becomes

$$\frac{d}{dt} u_m(t) \le (\gamma_0 + m\gamma) u_m(t) + f_m(t) + A_f \cdot 2C_0 (m-1)\gamma \cdot (k+m)!^s K_0(t) \left(K(t)C\right)^{k+m-1} \frac{t^{j+1}}{(j+1)!}.$$

This proves Lemma 4. In fact,

$$\int_0^t \exp[(\gamma_0 + m\gamma)(t-s)]\, f_m(s)\, ds \le A_f\, K_0(t) \left(K(t)C\right)^{k+m} \int_0^t \frac{s^j}{j!}\, ds.$$

$$\int_0^t \exp[(\gamma_0 + m\gamma)(t-s)]\, K_0(s) \left(K(s)C\right)^{k+m-1} \frac{s^{j+1}}{(j+1)!}\, ds$$

$$\le K_0(t) \left(\exp(\gamma t)C\right)^{k+m} C^{-1} \frac{t^{j+1}}{(j+1)!} \int_0^t (1+\gamma s)^{k+m-1}\, ds,$$

here the integral is estimated by $\frac{1}{(k+m)\gamma}(1+\gamma t)^{k+m}$, which is again estimated by $\frac{1}{m\gamma}(1 + \gamma t)^{k+m}$. Thus, the above integral is estimated by

$$\frac{1}{Cm\gamma} K_0(t) \left(K(t)C\right)^{k+m} \frac{t^{j+1}}{(j + 1)!} .$$

Since we assumed $C \geq 4C_0$ (see Lemma 4), we proved (16).

§ A.2 Proof of Lemmas 1,2 and 6.

Proof of Lemma 1. This inequality is well-known. Put

$$F(\rho) = \int |\xi|^{2\rho} |\hat{f}(\xi)|^2 \, d\xi.$$

Define $\Phi(\rho) = \frac{1}{2} \log F(\rho)$ ($= \log \| f \|_\rho$). Then the inequality becomes $\Phi(\rho) \leq (1 - \rho)\Phi(0) + \rho\Phi(1)$. This means that $\Phi''(\rho) \geq 0$. Now $\Phi''(\rho) \geq 0$ is equivalent to $F F'' - F'^2 \geq 0$. Since $F' = 2 \int \log|\xi| \cdot |\xi|^{2\rho} |\hat{f}(\xi)|^2 \, d\xi$, $F'' = 4 \int (\log|\xi|)^2 |\xi|^{2\rho} |\hat{f}(\xi)|^2 \, d\xi$, the last inequality follows from the Cauchy-Schwarz inequality. □

Proof of Lemma 2. It is enough to show that $a!^{1-\rho}(a+1)!^{\rho}/(a+\rho)!$, as a function of a, remains bounded when a tends to ∞, because this is a continuous function of a (≥ 0). Stirling's formula shows

$$a!^{1-\rho}(a+1)!^{\rho} \underset{a\to\infty}{\sim} \sqrt{2\pi a}^{\,1-\rho} \sqrt{2\pi(a+1)}^{\,\rho} \left(\frac{a}{e}\right)^{a(1-\rho)} \left(\frac{a+1}{e}\right)^{(a+1)\rho} .$$

$$(a + \rho)! \sim \sqrt{2\pi(a + \rho)} \left(\frac{a + \rho}{e}\right)^{a+\rho} .$$

its ratio $\underset{a\to\infty}{\sim} \left(a^{1-\rho}(a+1)^{\rho}/_{a+\rho}\right)^{\frac{1}{2}} \cdot a^{a(1-\rho)}(a+1)^{(a+1)\rho} \Big/ (a+\rho)^{a+\rho}$

$$\sim \left(1+\frac{1}{a}\right)^{(a+1)\rho} \Big/ \left(1+\frac{\rho}{a}\right)^{a+\rho} \sim 1. \quad \square$$

Proof of Lemma 6. By Leibniz,

$$\| \partial^{\alpha}(bv) \| \le ABC^{|\alpha|} \sum_{\alpha'} \binom{\alpha}{\alpha'} (k+|\alpha'|)!^{s} (k'+|\alpha-\alpha'|)!^{s}.$$

Since $\sum_{|\alpha'|=p} \binom{\alpha}{\alpha'} \le \binom{|\alpha|}{p}$, $\sum$ in right-hand side is estimated by

$$\sum_{0\le p\le|\alpha|} \binom{|\alpha|}{p} (k+p)!^{s}(k'+|\alpha|-p)!^{s} \le (k+k'+|\alpha|)!^{s} \sum_{p} \binom{|\alpha|}{p} \binom{k+k'+|\alpha|}{k+p}^{-s}.$$

Now $\binom{k+k'+|\alpha|}{k+p} \ge \binom{k+k'}{k}\binom{|\alpha|}{p}$, this proves Lemma 6. $\square$

§ A.3 Proof of Lemma 3.

The proof is fairly heavy.

(First step) Put

$$av = a\langle\Lambda\rangle^{-\rho}\left(\langle\Lambda\rangle^{\rho}v\right) = a_0\left(\langle\Lambda\rangle^{\rho}v\right)$$

where $\langle\xi\rangle = \left(1+|\xi|^2\right)^{\frac{1}{2}}$, and $\widehat{\langle\Lambda\rangle^{\rho}v} = \langle\xi\rangle^{\rho}\hat{v}(\xi)$.

Since $a_0(x,\xi) \in S^{0}_{0,0}$, by applying Calderón-Vaillancourt to $\partial_x^{\alpha} a_0(x,\xi)$ (see Theorem A.1, in Lecture III),

$$\left\| \partial_x^{\alpha} a_0(x,D) \right\|_{(L^2;L^2)} \leq c(a_0)\cdot|\alpha|!^s \, (2C_0)^{|\alpha|}, \quad \forall \alpha \geq 0 . \tag{A.7}$$

Next, by definition of $<\Lambda>^{\rho}$, there exists a constant $c_0(\rho)$, such that

$$\| \partial^{\alpha} <\Lambda>^{\rho} v \| \leq c_0(\rho)\left[\| \partial^{\alpha} v \|_{\rho} + \| \partial^{\alpha} v \| \right].$$

Applying Lemma 1 to $\| \partial^{\alpha} v \|_{\rho}$, we get ($\exists\, c_1(\rho)$),

$$\| \partial^{\alpha} <\Lambda>^{\rho} v \| \leq c_1(\rho)\left(\| \partial^{\alpha} v \| + \| \partial^{\alpha} v \|^{1-\rho} \| \partial^{\alpha} v \|_1^{\rho} \right) . \tag{A.8}$$

By Leibniz,

$$\partial^{\alpha}(av) = \sum_{\alpha' \leq \alpha} \binom{\alpha}{\alpha'} (\partial_x^{\alpha'} a_0)\left(\partial^{\alpha''} <\Lambda>^{\rho} v\right).$$

Using (A.8),

$$\| \partial^{\alpha}(av) \| \leq c_1(\rho) \sum_{\alpha'} \binom{\alpha}{\alpha'} \| \partial_x^{\alpha'} a_0 \| \left(\| \partial^{\alpha''} v \| + \| \partial^{\alpha''} v \|^{1-\rho} \| \partial^{\alpha''} v \|_1^{\rho} \right) .$$

We sum up all these inequalities for all α, $|\alpha| = m$, to obtain (taking account of (A.7)),

$$(av)_m \leq c(a_0)\, c_1(\rho)\, (I_1 + I_2), \quad \text{where}$$

$$I_1 = \sum_{|\alpha|=m} \sum_{\alpha'} \binom{\alpha}{\alpha'} |\alpha'|!^s (2C_0)^{|\alpha'|} \| \partial^{\alpha''} v \| ,$$

$$I_2 = \sum_{|\alpha|=m} \sum_{\alpha'} \binom{\alpha}{\alpha'} |\alpha'|!^s (2C_0)^{|\alpha'|} \| \partial^{\alpha''} v \|^{1-\rho} \| \partial^{\alpha''} v \|_1^{\rho} .$$

(<u>Second step</u>)

First consider I_1. <u>We fix α''</u>, and consider the estimate of the coefficient of $\| \partial^{\alpha''} v \|$. $\binom{\alpha}{\alpha'} \le \binom{|\alpha|}{|\alpha'|}$, and the number of $\{\alpha;\ \alpha \ge \alpha'',\ |\alpha| = m\}$ is evidently $\binom{m-|\alpha''|+n-1}{n-1}$ (Recall n is the number of components of $(\alpha_1, \cdots, \alpha_n)$). We estimate this by $2^{n-1+m-|\alpha''|}$. Thus, in I_1, the coefficient of $\| \partial^{\alpha''} v \|$ is estimated by

$$2^{n-1}\, 2^{m-|\alpha''|} \binom{m}{m-|\alpha''|} (m - |\alpha''|)!^s\, (2C_0)^{m-|\alpha''|}.$$

Denoting $|\alpha''| = p$, and observing $2^{m-|\alpha''|}\, (2C_0)^{m-|\alpha''|} = (4C_0)^{m-p} \le C^{m-p}$, we get

$$I_1 \le 2^{n-1} \sum_{0\le p\le m} \binom{m}{m-p} (m - p)!^s\, C^{m-p}\, v_p. \tag{A.9}$$

Since $v_p \le M(k + p)!^s\, C^p$,

$$I_1 \le 2^{n-1}\, MC^m \sum_{p} \binom{m}{m-p} (m - p)!^s\, (k + p)!^s.$$

Now, $\sum \cdots \le (k + m)!^s \sum_p \binom{m}{m-p} \binom{k+m}{k+p}^{-s}$. Taking account of $\binom{k+m}{k+p} \ge \binom{m}{p}$, this is estimated by $(k+m)!^s \sum_{0\le p\le m} \binom{m}{p}^{-s+1} \le (k+m)!^s\ c(s-1)$ (see (40)).

$$(A.10)$$

Next, we pass to I_2. We follow the above argument. For this purpose, observe that (Hölder's inequality),

$$\sum_{|\alpha''|=p} \| \partial^{\alpha''} v\|^{1-\rho} \| \partial^{\alpha''} v\|_1^{\rho} \le \left(\sum_{|\alpha''|=p} \| \partial^{\alpha''} v\| \right)^{1-\rho} \left(\sum_{|\alpha''|=p} \| \partial^{\alpha''} v\|_1\right)^{\rho}.$$

From this, we get instead of (A.9)

$$I_2 \le 2^{n-1} \sum_{0\le p\le m} \binom{m}{m-p} (m-p)!^s\, C^{m-p}\, v_p^{1-\rho}\, n^{\rho}\, v_{p+1}$$

$$\le 2^{n-1}\, n^{\rho} M C^{m+\rho} \sum_{0\le p\le m} \binom{m}{m-p} (m-p)!^s\, (k+p)!^{s(1-\rho)}\, (k+p+1)!^{s\rho}.$$

Now, by Lemma 2, $(k+p)!^{1-\rho}(k+p+1)!^{\rho} \le c(\rho)\cdot(k+p+\rho)!$. Thus,

$$I_2 \le \left(2^{n-1} c(\rho)^s\, n^{\rho}\right) M C^{m+\rho} \sum_p \binom{m}{m-p} (m-p)!^s\, (k+p+\rho)!^s.$$

By the same arguments as above for I_1, we obtain

$$I_2 \le \left(2^{n-1} c(\rho)^s\, n^{\rho} c(s-1)\right) M\, (k+m+\rho)!^s\, C^{m+\rho}. \tag{A.11}$$

Comparing (A.10) and (A.11), evidently $I_1 \le I_2$ (observe $c(\rho) \ge 1, C \ge 1$). Thus denoting

$$K_a = 2\, c(a_0)\, c_1(\rho) \left(2^{n-1}\, c(\rho)^s\, n^{\rho} c(s-1)\right), \tag{A.12}$$

we get

$$(av)_m \le K_a\, M\, (k+m+\rho)!^s\, C^{m+\rho}.$$

References

[1] M.S. Boutet de Monvel et P. Krée, Pseudo-differential operators and Gevrey classes, Ann. Inst. Fourier, 17(1967), 295-323.

[2] M.D. Bronshtein, The parametrix of the Cauchy problem for hyperbolic operator with characteristics of variable multiplicity, Trudy Moscow Math., 41(1980), 83-99.

[3] J. Chazarain, Opérateurs hyperboliques à caract ristiques de constante multiplicité, Ann. Inst. Fourier, 17(1974), 173-202.

[4] Y. Demay, Paramétrix pour des systèmes hyperboliques du premier ordre à multiplicité constante, J. Math. pure et appl., 56(1977), 393-422.

[5] J. De Paris, Problème de Cauchy analytique à données singulières pour un opérateur différentiel bien décomposable, J. Math. pure et appl., 51(1972), 465-88.

[6] H. Flaschka and G. Strang, The correctness of the Cauchy problems, Adv. in Math., 6(1971), 349-79.

[7] J. Hadamard, Le probl me de Cauchy, Hermann (1932).

[8] S. Hashimoto-T. Matsuzawa-Y. Morimoto, Opérateurs pseudo-différentiels et classes de Gevrey, Comm. in P.D.E., 8(1983), 1277-89.

[9] L. Hörmander, Linear partial differential operators, Springer-Verlag (1963).

[10] V. Ivrii, Correctness of the Cauchy problem in Gevrey class for non-strictly hyperbolic operators, Mat. Sbornik, 96(1975), 365-87.

[11] K. Kajitani, Leray-Volevich's systems and Gevrey classes, J. Math. Kyoto Univ., 21(1981), 547-74.

[12] ——————, Cauchy problem for nonstrictly hyperbolic systems in Gevrey classes, J. Math. Kyoto Univ., 23(1983), 599-616.

[13] ——————, Cauchy problems for non-strictly hyperbolic systems, Publ. Res. Inst. Math. Sci., 15(1979), 519-50.

[14] H. Komatsu, Linear hyperbolic equations with Gevrey coefficients, J. Math. pure et appl., 59(1980), 145-85.

[15] H. Kumano-go, Pseudo-differential operators, M.I.T. Press (1981).

[16] J. Leray et Y. Ohya, Systèmes linéaires, hyperboliques nonstricts, Deuxième Colloque sur l'Analyse fonctionnelle à Liège, 1964.

[17] ——————————, Systèmes nonlinéaires hyperboliques nonstricts, Math. Ann. 70(1967), 167-205.

[18] W. Matsumoto, On the conditions for hyperbolicity of systems with double characteristic roots, I, II, J. Math. Kyoto Univ., 21(1981), 47-84, 251-71.

[19] S. Mizohata, On the hyperbolicity in the domain of real analytic functions and Gevrey classes, Hokkaido Math. J., 12(1983), 298-310.

[20] T. Nishitani, On the Lax-Mizohata theorem in the analytic and Gevrey classes, J. Math. Kyoto Univ., 18(1979), 501-12.

[21] ————————, Cauchy problem for nonstrictly hyperbolic systems in Gevrey classes, J. Math. Kyoto Univ., 23(1983), 739-73.

[22] Y. Ohya, Le problème de Cauchy pour les équations hyperboliques à caractéristique multiple, J. Math. Soc. Japan, 16(1964), 268-86.

[23] —————, Le problème de Cauchy à caractéristiques multiples, Cours à l'Univ. de Paris IV, (1981).

[24] F. Treves, Introduction to pseudo-differential and Fourier integral operators, 1, 2, (1980), Plenum Press.

[25] J. Vaillant, Systèmes hyperboliques à caractéristiques multiples, J. Math. pure et appl., 58(1979), 165-216.

[26] H. Yamahara, On the Cauchy problem for weakly hyperbolic systems, Publ. Res. Inst. Math. Sci., 12(1976), 493-512.

[27] V.Ya. Ivrii, Conditions for correctness in Gevrey classes of the Cauchy problem for weakly hyperbolic equations, Siberian J. of Math.,17(1976), 422-35, (English translation).

[28] V.M. Petkov, On the Cauchy problem for first order hyperbolic systems with multiple characteristics, Soviet Math. Dokl., 14(1974), 534-37.

[29] S. Mizohata, Sur l'indice de Gevrey, à paraître au Séminaire sur les équations aux dérivées partielles de J. Vaillant(1983-84), Hermann.

Lecture V. Micro-local Analysis in Gevrey Class (I).

§ 1. Introduction.

In last lecture we showed that the treatment of the non-strict hyperbolic equation with constant multiplicity can be reduced to the study of the elementary operator. To be precise, the essential part of the argument was mainly concerned with the elementary operator,

$$L(u) \equiv \partial_t u + i\lambda(x,t;D)\, u + c(x,t;D)\, u = f, \tag{1}$$

where order $c = \rho$, with $\rho < 1$.

Let us repeat the following facts: If $\rho \le 0$, the Cauchy problem with $u\big|_{t=0} = u_0(x)$ is H^{∞}-wellposed. If $\rho > 0$, let $c(x,t;\xi) = \mathring{c}(x,t;\xi) + \tilde{c}(x,t;\xi)$ where $\mathring{c}(x,t;\xi)$ is the (essentially)homogeneous part of degree ρ, and $\tilde{c}$ is of order ≤ 0, then in the case when $\operatorname{Re}\mathring{c}(x,t;\xi) \ge 0$, (1) is still H^{∞}-wellposed. Contrary to this, if there exists an (x_0, ξ^0) such that

$$\operatorname{Re}\mathring{c}(x_0,0;\xi^0) < 0,$$

then the forward Cauchy problem is no longer H^{∞}-wellposed. However if $\frac{1}{s} > \rho$, the Cauchy problem is well-posed in $\gamma^{(s)}$, assuming, of course, $\lambda(x,t;\xi)$, $c(x,t;\xi)$ and $f(x,t)$ are in Gevrey class $\gamma^{(s)}$. Moreover, although we did not prove, we can show that $s \le \frac{1}{\rho}$ is necessary.

We want to show the micro-local version of Lecture IV. What is the definition of the micro-local analysis ? It seems that there is no precise definition of it. It is not to be wondered at. This was the same with " Functional Analysis ". The author has the following experience. One day, when the author was staying in Chicago in 1960, Zygmund, Calderon and several mathematicians were talking about functional analysis. Then, Zygmund proposed to make precise the definition of functional analysis. (Maybe, he did it with the singular integral operator in mind.) Several subjects were enumerated without arriving at a satisfactory answer. Finally Calderon said, " There is no definite definition. However we can make fairly concrete image, looking at the book by Riesz and Nagy ". Recall that in the study of P.D.E., in the fifties and in the first half of the sixties, the global existence theorem, by using rich results in functional analysis was in fashion. Nowadays, the micro-local analysis is becoming fashionable analysis. The new name (or new face) attracts young mathematicians.

Now we return to today's talk. The author is thinking that the contents of Lecture III give already some image to the micro-local analysis. However the author has some hesitation to call the contents of Lecture III, V, VI as micro-local analysis, because those are quite different from the one which Sato school calls micro-local analysis. They develop systematically the cohomological arguments for pseudo-differential operators with analytic coefficients. In that sense, to distinguish ours from them, it would be safe to call our analysis as <u>local Fourier analysis</u> or <u>micro-local treatment</u>.

The starting point of the micro-local analysis may be the notion of wave front set, denoted by WF(u). This notion was introduced by Hörmander and Sato. The name <u>wave front set</u> is not appropriate. Several mathematicians begin to call this <u>singular spectrum</u>. Anyway, name is name. The content is important. Here we call wave front set. Let us explain this.

In this lecture we denote the dimension of the x-space by ℓ instead of n (as in Lecture III), because we use micro-localizer $\alpha_n(D)$. Let $u(x)$ be a distribution defined in an open set Ω ($\subset \mathbb{R}^\ell$). Let $x_0 \in \Omega$. Take a cut-off function $\zeta(x) \in C_0^\infty$ with small support around x_0, which takes the value 1 in a neighborhood of x_0. Take the Fourier transform $\widehat{\zeta u}(\xi)$. We know that if $u(x)$ is C^∞ in a neighborhood of x_0, then taking the support of ζ small, $\widehat{\zeta u}(\xi)$ decreases faster than any inverse power of $|\xi|$ when $|\xi|$ tends to ∞. Namely, for any positive integer N, there exists a constant C_N such that

$$|\widehat{\zeta u}(\xi)| \le C_N(1 + |\xi|)^{-N}. \tag{2}$$

The converse is also true. Namely, if (2) holds by choosing suitable cut-off function $\zeta(x)$, then $u(x)$ is C^∞-function in a neighborhood of x_0. For general $u(x)$, $|\widehat{\zeta u}(\xi)|$ is of at most polynomial growth in $|\xi|$. Now, ξ^0 ($|\xi^0| = 1$) is fixed, and we look at $\widehat{\zeta u}(\xi)$ for $\xi = \rho\xi^0$ ($\rho > 0$, tending to ∞). Then, we could say that if (2) holds, ξ^0 is a good direction, and if (2) fails to hold, ξ^0 is a bad direction. We are interested in the bad direction from the view-point of the analysis of singularities of u. However it is sometimes difficult to decide this. Taking account of this, we slightly change the above view-point. We could say ξ^0 is a bad direction, if for any small conic neighborhood of ξ^0 (see the Figure), (2) fails to hold. In this definition, $\zeta(x)$ is fixed. To make this notion more complete, we should also consider a family of $\zeta(x) \in C_0^\infty$, satisfying above condition. To be precise, ξ^0 is a bad direction if for any choice of $\zeta(x)$, ξ^0 is a bad direction.

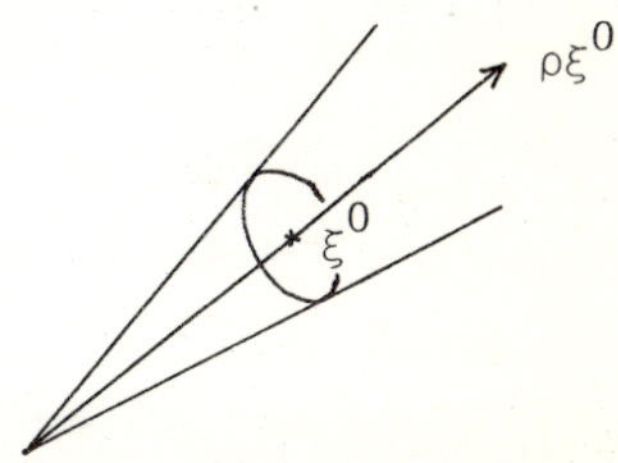

The definition of the wave front set $WF(u)$ is given in the following way.

$$
\begin{cases}
(x_0, \xi^0) \notin WF(u), \text{ if and only if there exist a } \zeta(x) \in C_0^\infty \text{ taking the} \\
\text{value } 1 \text{ in a neighborhood of } x_0, \text{ and a conic neighborhood } V_{\xi^0} \text{ of} \\
\xi^0, \text{ such that (2) holds for } \xi \in V_{\xi^0}.
\end{cases} \tag{3}
$$

This notion seems to be not entirely new. However this notion makes the considerations of the singularities or the regularities of solutions much easier (see Proposition 1 and 2 in § 4 and § 5). We can say more. The fruitfulness of this notion is revealed by the following fact (due to Hörmander).

Singularities propagate along bicharacteristic strips.

Let us explain this for solutions of (1). Take a point x_0 in the initial hyperplane $t = 0$, and take a direction ξ^0 ($\in \mathbb{R}^\ell \setminus 0$). Let $(x(t), \xi(t))$ be the bicharacteristic strip issued from (x_0, ξ^0) at $t = 0$. Namely, $(x(t), \xi(t))$ is the solution of the Hamiltonian flow:

$$
\begin{cases}
\dfrac{dx}{dt} = \lambda_\xi(x, t; \xi) \\
\\
\dfrac{d\xi}{dt} = -\lambda_x(x, t; \xi)
\end{cases}
$$

$$(x(0), \xi(0)) = (x_0, \xi^0).$$

Suppose order $c \leq 0$ in (1). Then

Fundamental Theorem .

If $(x_0, \xi^0) \notin (\in)\ WF(u(\cdot,0))$, then it follows $(x(t), \xi(t)) \notin (\in)$ $WF(u(\cdot,t))$.

In this lecture and the next one, we want to show that <u>the result is also true in Gevrey class</u> $\gamma^{(s)}$ $(s > 1)$. For this, let us explain the wave front set in the sense of Gevrey $\gamma^{(s)}$, denoted by $WF_s(u)$.

If $u(x) \in \gamma^{(s)}$ with compact support, then $(\exists \varepsilon_0 > 0)$

$$|\hat{u}(\xi)| \leq \exp(-\varepsilon_0|\xi|^{\frac{1}{s}}), \quad \text{if } |\xi| \text{ is large.} \tag{5}$$

Conversely, if $\hat{u}(\xi)$satisfies (5), then it is easily seen that $u(x) \in \gamma^{(s)}$. Taking account of this, we say that $(x_0, \xi^0) \in WF_s(u)$, if and only if there exists a cut-off function $\zeta(x) \in \gamma^{(s)}$ with compact support, taking the value 1 in a neighborhood of x_0 such that the estimate

$$|\widehat{\zeta u}(\xi)| \leq \exp(-\varepsilon_0|\xi|^{\frac{1}{s}}) \quad (\exists\, \varepsilon_0 > 0) \tag{6}$$

holds, when ξ tends to ∞, remaining in a suitable conic neighborhood V_{ξ^0} of ξ^0.

We consider always equation (1). Suppose all the coefficients and $f(x,t)$ belong to $\gamma_x^{(s)}$. Suppose further

$$\text{order } c(x,t;\xi) = \rho < \frac{1}{s}. \tag{7}$$

Then,

<u>Fundamental Theorem in Gevrey class</u>.

If $(x_0, \xi^0) \notin (\in)\ WF_s(u(\cdot,0))$, then $(x(t),\xi(t)) \notin (\in)\ WF_s(u(\cdot,t))$. (8)

It seems to the author, the treatment in Gevrey class cannot be done by a straightforward extension of the C^∞-case. Let us explain a standard method in C^∞-case. (see, Nirenberg [6]). If we construct the micro-localizer $K = k(x,t;D)$ such that

$$[k, L] \sim 0 \quad (\sim \text{ means " modulo regularizing operator "}),$$

$$k(x,0;\xi) = \alpha(\xi)\zeta(x),$$

where $\alpha(\xi)$ is essentially homogeneous of degree 0 with its support in a conic neighborhood of ξ^0, then

$$K(Lu) = L(Ku) + [K, L]\, u,$$

which implies

$$L\big(k(x,t;D)u(\cdot,t)\big) = g \in C_x^\infty.$$

Now we know that if $k(x,0;D)u(\cdot,0) \in C^\infty$, then for all t, $k(x,t;D)u(\cdot,t) \in C^\infty$.

The equation $[k, L] \sim 0$ can be solved by successive approximations, taking account of the asymptotic formula for pseudo-differential operators. In fact, putting

$$k(x,t;\xi) \sim k_0 + k_{-1} + k_{-2} + \cdots ,$$

k_0 should satisfy

$$\begin{aligned} &\{\partial_t + \lambda_\xi(x,t;\xi)\partial_x - \lambda_x(x,t;\xi)\partial_\xi\} k_0(x,t;\xi) = 0 \\ &k_0(x,0;\xi) = \alpha(\xi)\zeta(x). \end{aligned} \tag{9}$$

This equation is called the transport equation. Now, in the case of Gevrey $\gamma^{(s)}$, if we could construct K such that

$$[k, L] \sim 0 \text{ in } \gamma^{(s)},$$

the proof would be complete. However, it seems to be almost impossible to expect this. We can say merely that $[K, L] \sim 0$ in $\gamma^{(2s-1)}$. This is the case with Friedmann (1962) [1] and Kajitani [3]. Recently, Taniguchi succeeded in proving (8) by the consideration of the multi-phase products of Fourier integral operators [7]. His method is fairly complicated. Our method is the so-called (α, β)-method explained in Lecture III. We adopt this method to the analysis in Gevrey class. For this purpose, we prepare several facts, especially we show a criterion of $WF_s(u)$.

§ 2. Definition of $\{\alpha_n(D), \beta_n(x)\}$.

We used already a kind of micro-localizer $\{\alpha_n(D), \beta(x)\}$ in the study of the wellposedness in Lecture III. Here we use the same micro-localizer. However, in the treatment of Gevrey class, we need to impose more strict restrictions on $\{\alpha_n(\xi), \beta(x)\}$. Our actual subject is clearly connected with the work of Hörmander [2]. In this paper he proved for the differential operator $P(x, \partial_x)$ with analytic coefficients that if (x_0, ξ^0) is a simple characteristic point of real $P_m(x, \xi)$ (principal symbol of P), then $WF_A(u)$ of $P u = f$ with analytic f propagates along the null-bicharacteristic strip through (x_0, ξ^0).

We use the following cut-off function which was used by Hörmander (Egorov also used similar functions).

<u>Definition of $\alpha_n(\xi)$.</u> Let $\chi_N(\xi)$ be a C_0^∞-function depending on parameter N, having its support in $|\xi - \xi^0| \le r_0$, $(0 \le \chi_N(\xi) \le 1)$, satisfying the following two conditions:

$$\left\{\begin{array}{ll} \chi_N(\xi) = 1 \quad \text{for} \quad |\xi - \xi^0| \le r_0/2, & \\ |\chi_N^{(\mu+\nu)}(\xi)| \le \left(Ncr_0^{-1}\right)^{|\mu|} \nu!^{1+\varepsilon'} \left(cr_0^{-1}\right)^{|\nu|} & \text{for} \quad |\mu| \le N \end{array}\right. \tag{10}$$

$$(s > 1, \quad 1 + 2\varepsilon' < s, \quad \varepsilon' > 0),$$

where c can be considered as an absolute constant (namely, independent of r_0). Let us remark that ν can be taken arbitrarily, but μ is restricted by $|\mu| \le N$. If $\nu = 0$, then this estimate can be considered as an analytic estimate, for in the case $|\mu| = N$, $\left(Ncr_0^{-1}\right)^N \sim (2\pi N)^{-1/2}\left(ecr_0^{-1}\right)^N N!$.

Put

$$\alpha_n(\xi) = \chi_N(\xi/n), \tag{11}$$

therefore

$$\left|\alpha_n^{(\mu+\nu)}(\xi)\right| \le \left(\frac{N}{n}\, cr_0^{-1}\right)^{|\mu|} \nu!^{1+\varepsilon'} . \left(cr_0^{-1}\right)^{|\nu|} \Big/ n^{|\nu|}, \quad \text{for } |\mu| \le N.$$

Here n and N are related by

$$N = \theta\, n^{\frac{1}{s}}, \quad \theta = (ce)^{-1}\, r_0. \tag{12}$$

<u>Definition of $\beta_n(x)$</u>. $\beta_n(x)$ ($\in C_0^\infty$) ($0 \le \beta_n(x) \le 1$) has its support in $|x - x_0| \le r_0$, and $\beta_n(x) = 1$ for $|x - x_0| \le {}^{r_0}/_2$. Moreover

$$|\beta_{n\,(\mu+\nu)}(x)| \le \left(Ncr_0^{-1}\right)^{|\mu|} \nu!^{1+\varepsilon'} \left(cr_0^{-1}\right)^{|\nu|}, \quad \text{for } |\mu| \le N, \tag{13}$$

where N is the same integer defined by (12).

<u>Note</u>. In the above definition, N is an integer. In that sense, (12) is not correct. It is better to say that N is the nearest integer to $\theta\, n^{1/s}$. However we are concerned with large n, thus there is no serious danger to write this in the form (12). The above conditions on $\alpha_n(\xi)$ and $\beta_n(x)$ enable us to use two different kinds of estimates. This makes our treatment fairly easy in the case of $\gamma^{(s)}$ with $s > 1$. Especially this is convenient for $\beta_n(x)$. Incidentally, in the treatment of the analytic wave front sets, we impose only

$$\begin{cases} |\alpha_n^{(\mu)}(\xi)| \le \left(\frac{N}{n} c r_0^{-1}\right)^{|\mu|}, & \text{for } |\mu| \le 2N \\ |\beta_{n_{(\nu)}}(x)| \le \left(N c r_0^{-1}\right)^{|\nu|}, & \text{for } |\nu| \le 2N \end{cases} \tag{14}$$

where $N = \frac{1}{2}(ce)^{-2} \ell^{-1} r_0^{\,2}$. □

Now, we want to express the condition: $(x_0, \xi^0) \notin WF_s(u)$ by means of $\{\| \alpha_n^{(p)} \beta_{n_{(q)}} u \| \}_{0 \le |p+q| \le N}$. First, the above condition says that there exists a $\zeta(x) \in \gamma^{(s)} \cap C_0^\infty$ $(0 \le \zeta(x) \le 1)$, which takes the value 1 in a neighborhood of x_0, and a conic neighborhood V_{ξ^0} of ξ^0 such that

$$|\widehat{\zeta u}(\xi)| \le \exp\left(- \varepsilon_1 |\xi|^{\frac{1}{s}}\right) \qquad (\exists \varepsilon_1 > 0) \tag{15}$$

when $\xi \in V_{\xi^0}$ and $|\xi|$ is sufficiently large. Thus if r_0 is small, then supp$[\alpha_n(\xi)]$ is contained in V_{ξ^0}, which implies

$$\left|\alpha_n(\xi)\ \widehat{\zeta u}(\xi)\right| \le \exp\left(- \varepsilon_1 \left((1 - r_0)n\right)^{\frac{1}{s}}\right).$$

This implies again, by Plancherel, the same kind estimate for $\| \alpha_n(D)\ \zeta u \|$. However, as in Lecture III, we are mainly concerned with the micro-localized function (solution) of the form $\alpha_n(D)\beta_n(x)\,u$. Thus in order to reflect the decreasing order of $|\widehat{\zeta u}(\xi)|$ on this function, we must commute $\alpha_n(D)$ with $\beta_n(x)$. To be precise, let us assume $\beta_n \Subset \zeta$ (<u>this means that, on the support of</u> $\beta_n(x)$, $\zeta(x) = 1$). This is always realized in our arguments, because in our consideration, we need almost always to take r_0 (which determines the size of $\alpha_n(\xi)$, $\beta_n(x)$) small. Then

$$\alpha_n(D)\beta_n(x)\, u = \alpha_n(D)\beta_n(x)\zeta(x)\, u \sim \sum_{\nu\geq 0} \nu!^{-1}\beta_{n_{(\nu)}}\alpha_n^{(\nu)}\zeta u.$$

In view of application, in the treatment of $\gamma^{(s)}$ with $s > 1$, we put

$$S_n = \sum_{|p+q|\leq N} c_{pq}^n \| \alpha_n^{(p)}\beta_{n_{(q)}} u \| ,$$

$$c_{pq}^n = n^{(1-1/s)|p|-1/s|q|}, \tag{16}$$

where N is defined by (12).

The reader might understand the necessity of such an energy form in view of Lecture III. The reason that we defined N by(12)is partly because of the fact that (we use it very often)

$$c_{pq}^n |\alpha_n^{(p)}|_0 \cdot |\beta_{n_{(q)}}|_0 \leq n^{(1-1/s)|p|-1/s|q|} \left(\tfrac{N}{n}\, cr_0^{-1}\right)^{|p|} \left(Ncr_0^{-1}\right)^{|q|}$$

$$= \left(Ncr_0^{-1} \Big/ n^{1/s}\right)^{|p+q|} = e^{-|p+q|},$$

and this becomes e^{-N} ($= \exp(-\theta n^{1/s})$) when $|P + q| = N$.

Finally, we assume here $u \in (\gamma^{(s)})'$ <u>with compact support</u>. u is no longer distribution, but ultradistribution. It is known that $\hat{u}(\xi)$ is defined and it has the following type estimate (see Komatsu [4]):

$$|\hat{u}(\xi)| \leq C_\varepsilon \exp\left(\varepsilon|\xi|^{\frac{1}{s}}\right) \quad (^\forall \varepsilon > 0, \ ^\exists C_\varepsilon). \tag{17}$$

The fundamental solution of the Cauchy problem is no longer distribution, this is ultradistribution.

§ 3. Criterion of $WF_s(u)$ by S_n.

Theorem 1. Suppose $(x_0, \xi^0) \notin WF_s(u)$ (with $|\xi^0| = 1$), $u \in (\gamma^{(s)})'$ with compact support. Then we have, if r_0 is small,

$$S_n = \sum_{|p+q| \le N} c^n_{pq} \| \alpha_n^{(p)} \beta_{n_{(q)}} u \| \le \exp\left(-\varepsilon n^{\frac{1}{s}}\right)$$

if n is large, where ε is a positive constant which we can choose as the one independent of r_0 when r_0 tends to 0.

The converse is also true. We can state it in a strong form.

Theorem 2. Suppose $u \in (\gamma^{(s)})'$ with compact support. Then an estimate of the form

$$\tilde{S}_n = \sum_{|p| \le N} c^n_{p0} \| \alpha_n^{(p)} \beta_n u \| \le \exp\left(-\varepsilon_0 n^{\frac{1}{s}}\right) \qquad (\exists \varepsilon_0 > 0),$$

for n large, implies $(x_0, \xi^0) \notin WF_s(u)$.

The proof of Theorem 1 is a little delicate, because we need to consider a partition of unity in the ξ-space. In the following, we give the proof of Theorem 2. First, we take a $\zeta(x) \in C_0^\infty$ such that on its support $\beta_n(x) = 1$ (i.e. $\zeta \Subset \beta_n$). More precisely, $\mathrm{supp}[\zeta] = \{x; |x - x_0| \le \frac{r_0}{2}\}$, $\zeta(x) = 1$, for $|x - x_0| \le \frac{r_0}{4}$. We may assume

$$|\zeta_{(\nu)}(x)| \leq \nu!^{1+\varepsilon'} \left(2cr_0^{-1}\right)^{|\nu|}.$$

We can show

$$\|\alpha_n \zeta u\| \leq \exp\left(-\varepsilon_0' n^{\frac{1}{s}}\right) \qquad (\exists \varepsilon_0' > 0) \tag{18}$$

if n is large. Its proof is given in Appendix.

Next, consider $\alpha(\xi) \in C^\infty$ $(0 \leq \alpha(\xi) \leq 1)$, whose support is contained in a small conic neighborhood of ξ^0, taking the value 1 in a smaller conic neighborhood of ξ^0. We assume also $\mathrm{supp}[\alpha(\xi)] \subset \{\xi;\ |\xi| \geq n_0\}$ (n_0 is large).

$$|D^m\alpha(D)\zeta u|_0 \leq (2\pi)^{-\ell}\iint |\xi|^m|\alpha(\xi)\,\widehat{\zeta u}(\xi)|\, d\xi$$

$$\leq (2\pi)^{-\ell} \sum_{n=n_0}^{\infty} (2n)^m \iint \alpha_n(\xi)|\widehat{\zeta u}(\xi)|\, d\xi,$$

since on the support of $\alpha_n(\xi)$, $|\xi| \leq 2n$, and $\sum_{n=n_0}^{\infty} \alpha_n(\xi) \geq \alpha(\xi)$.

Now, by Schwarz, the last is estimated by

$$\text{const.} \sum_n (2\pi)^{m+\frac{\ell}{2}} \|\alpha_n \zeta u\| \leq \text{const.} \sum_n (2n)^{m+\frac{\ell}{2}} \exp\left(-\varepsilon_0' n^{\frac{1}{s}}\right)$$

$$\leq \text{const.} \left(m + \tfrac{\ell}{2}\right)!^s\, c_1^{m+\ell/2} \qquad (\exists c_1 > 0)$$

$$\leq \text{const.}\ m!^s\, c_2^m.$$

We can say more:

$$|\alpha(\xi)\widehat{\zeta u}(\xi)| \leq c_2 \exp\left(-\varepsilon\,|\xi|^{\frac{1}{s}}\right) \qquad (\exists c_2, \varepsilon > 0). \tag{19}$$

In fact, this is equivalent to showing that

$$|\alpha_n(\xi)\widehat{\zeta u}(\xi)| \le c_2 \exp\left(-\varepsilon' n^{\frac{1}{s}}\right) \qquad (\exists\, \varepsilon' > 0). \tag{20}$$

Now,

$$|\alpha_n(\xi)\widehat{\zeta u}(\xi)| \le \|\alpha_n(D)\,\zeta u\|_{L^1} \le c\|(1+|x|^2)^{\ell}\alpha_n(D)\,\zeta u\|_{L^2}.$$

Since

$$(1+|x|^2)^{\ell}\alpha_n(\zeta u) = \sum_{|\kappa|\le 2\ell} (-1)^{|\kappa|}\kappa!^{-1}\alpha_n^{(\kappa)}\left[(1+|x|^2)^{\ell}\right]_{(\kappa)}(\zeta u)$$

$$= \sum_{|\kappa|\le 2\ell} \alpha_n^{(\kappa)}\,\zeta_\kappa u,$$

where $\zeta_\kappa = (-1)^{|\kappa|}\kappa!^{-1}\left[(1+|x|^2)^{\ell}\right]_{(\kappa)}\zeta(x)$. Thus

$$|\alpha_n(\xi)\widehat{\zeta u}(\xi)| \le \sum_{|\kappa|\le 2\ell} \|\alpha_n^{(\kappa)}\zeta_\kappa u\|.$$

By the same arguments as in the proof of (18), we get

$$\|\alpha_n^{(\kappa)}\zeta_\kappa u\| \le \exp\left(-\varepsilon_1' n^{\frac{1}{s}}\right) \qquad (\exists\, \varepsilon_1' > 0).$$

These estimates will imply (20).

§ 4. Some comments on WF(u).

Concerning WF(u), we have the same kind of theorems as Theorem 1 and 2. Let us explain this briefly. In this case, $\alpha_n(\xi)$ and $\beta(x)$ were defined in Lecture II (Appendix), and in Lecture III. To be precise, $\alpha(\xi) \in C_0^\infty$, $0 \le \alpha(\xi) \le 1$, with support $\{\xi;\ |\xi - \xi_0| \le r_0\}$, and $\alpha(\xi) = 1$, for $|\xi - \xi_0| \le r_0/2$. $\alpha_n(\xi) = \alpha(\xi/n)$; $\beta(x) \in C_0^\infty$, $0 \le \beta(x) \le 1$, with support $\{x;\ |x - x_0| \le r_0\}$, and $\beta(x) = 1$ for $|x - x_0| \le r_0/2$.

We define

$$S_{n,k} = \sum_{|p+q|\le k} n^{\theta|p|-\theta|q|} \| \alpha_n^{(p)} \beta_{(q)} u \| ,$$

where $0 < \theta < 1$.

For convenience, we say a sequence a_n (≥ 0) is <u>rapidly decreasing</u> if for any large number M, a_n is estimated by n^{-M} when n tends to ∞.

<u>Theorem 3.</u> Suppose $(x_0, \xi^0) \notin WF(u)$. Then, if r_0 is small, $S_{n,k}$ is rapidly decreasing for any fixed k.

Put

$$\overset{\circ}{S}_{n,k} = \sum_{|p|\le k} n^{\theta|p|} \| \alpha_n^{(p)} \beta u \| \qquad (0 < \theta < 1).$$

<u>Theorem 4.</u> If, for some choice of $\{\alpha_n, \beta\}$ and θ, $\overset{\circ}{S}_{n,k}$ is rapidly decreasing for any fixed k, then $(x_0, \xi^0) \notin WF(u)$.

The proof is left to the reader.

As an application of the above theorems, we show the local version of the hypo-ellipticity (more precisely, hypo-ellipticity in x) of parabolic operators.

Let $a(x,t;D)$ be a differential operator of order m (≥ 1). Suppose for its principal symbol $a_m(x,t;\xi)$:

$$\text{At the point } (x_0, \xi^0), \quad |\xi^0| = 1,$$

$$\operatorname{Re} a_m(x_0,t;\xi^0) \leq -\delta, \qquad ({}^{\exists}\delta > 0, \text{ constant}, \quad t \in [0, T]). \tag{21}$$

We are concerned with a local solution $u(x,t)$ of

$$\partial_t u = a(x,t;D)\, u + f(x,t). \tag{22}$$

<u>Proposition 1.</u> Suppose $t \mapsto f(x,t)$ is continuous with values in $\mathcal{D}'_x$, and $(x_0, \xi^0) \notin WF(f(\cdot,t))$, ${}^{\forall}t \geq 0$. Then,

$$^{\forall}t > 0, \quad (x_0, \xi^0) \notin WF(u(\cdot,t)).$$

<u>Proof</u>. The argument which follows is entirely similar to the one which was used in the proof of Theorem 3 in Lecture III. So, we admit several points which we proved there.

First, we can assume $t \mapsto u(\cdot,t)$ is continuous with values in $\mathcal{D}'_x$ (partial-hypoellipticity in t). Thus taking $\zeta(x) \in C_0^\infty$, taking the value 1 in a neighborhood of x_0, and considering $\zeta(x)u(\cdot,t)$ instead of $u(\cdot,t)$, we can assume that

$$|\hat{u}(\xi,t)| \leq C(1+|\xi|)^L \quad (\exists L>0), \text{ for } t \in [0,T]. \tag{23}$$

As in Lecture III, we consider

$$\partial_t(\alpha_n{}^{(p)}\beta_{(q)}u) = a(\alpha_n{}^{(p)}\beta_{(q)}u) + \alpha_n{}^{(p)}\beta_{(q)}f + g_{n,p,q}.$$

By hypothesis, we have the following type energy inequality:

$$\frac{d}{dt}\|\alpha_n{}^{(p)}\beta_{(q)}u(\cdot,t)\| \leq -\delta' n^m\|\alpha_n{}^{(p)}\beta_{(q)}u(\cdot,t)\| + \|\alpha_n{}^{(p)}\beta_{(q)}f\| + \|g_{n,p,q}\|$$

$g_{n,p,q}$ is $[\alpha_n{}^{(p)}\beta_{(q)}, a]\,u$. Thus

$$g_{n,p,q} = \sum_{1\leq|\mu+\nu|\leq N-(p+q)} \mu!^{-1}\nu!^{-1}(-1)^{|\nu|}a^{(\nu)}_{(\mu)}(\alpha_n{}^{(p+\mu)}\beta_{(q+\nu)}u) + r_{N,p,q}\,u,$$

where in view of (23), and a result in Lecture III,

$$\|r_{N,p,q}\,u\| \leq n^{m-N+L}, \quad \text{if } n \text{ is large.}$$

Now p (positive integer) is given, we take N in such a way that

$$m - \frac{1}{2}N + L = -p, \quad \text{i.e.} \quad N = 2(m+L+p).$$

We define,

$$S_{n,N}(t;u) = \sum_{|p+q|\leq N} n^{\frac{1}{2}|p|-\frac{1}{2}|q|}\|\alpha_n{}^{(p)}\beta_{(q)}u(\cdot,t)\|.$$

Now,

$$\begin{aligned} &n^{\frac{1}{2}(|p|-|q|)}\|g_{N,p,q}\| \\ &\quad\leq C(N)\, n^m \sum n^{-\frac{1}{2}|\mu+\nu|}\left(n^{\frac{1}{2}|p+\mu|-\frac{1}{2}|q+\nu|}\|\alpha_n{}^{(p+\mu)}\beta_{(q+\nu)}u(\cdot,t)\|\right) \\ &\quad + \text{const. } n^{m-\frac{1}{2}N+L}. \end{aligned}$$

This implies

$$\sum_{|p+q|\le N} n^{\frac{1}{2}(|p|-|q|)} \| g_{N,p,q} \| \le \text{const. } C(N)\, n^m S_{n,N} \Big/ n^{\frac{1}{2}} + O(n^{-p}).$$

Thus, for n large, we have

$$\frac{d}{dt} S_{n,N}(t;u) \le -\delta'' n^m S_{n,N}(t;u) + S_{n,N}(t;f) + O(n^{-p}) \quad (\ \delta'' > 0\ ).$$

By hypothesis (in view of Theorem 3), $S_{n,N}(t;f)$ is uniformly rapidly decreasing. Hence this term is absorbed in $O(n^{-p})$.

Now, we see that, by (23), $S_{n,N}(0;u) = O(n^L)$. Thus the above differential inequality gives

$$S_{n,N}(t;u) \le \exp\left(-\delta'' n^m t\right) O(n^L) + O(n^{-p}).$$

This implies, for all $t > 0$, $S_{n,N}(t;u) = O(n^{-p})$. Since p is arbitrary, by Theorem 4, we proved Proposition. □

Note. Proposition proves the hypo-ellipticity of parabolic operators. In fact the differentiability in x of u follows from Proposition and its differentiability in t follows if we assume the differentiability in t of $a(x,t;\xi)$ and of $f(x,t)$. Proposition proves also the hypo-ellipticity of elliptic operators. In fact, let $a(x,D)$ be an elliptic operator. If necessary, considering $\bar{a}(x,D)a(x,D)$ instead of a, we can assume (21) $\forall (x_0, \xi^0)$. Now, a solution $au = f$ satisfies (22) (Observe $\partial_t u = 0$).

§ 5. Some comments on $WF_A(u)$.

The definition of the analytic wave front set is delicate(see Hörmander [2]). Let us explain it briefly. First we suppose $u(x)$ to be real-analytic in a neighborhood of x_0 ($x_0 \in \Omega \subset \mathbb{R}^{\ell}$). To be precise,

$$|\partial^{\alpha} u(x)| \le C^{1+|\alpha|}\alpha!, \quad ({}^{\exists} C > 0), \quad \text{for} \quad |x - x_0| \le r_0.$$

We consider the following sequence of cut-off functions $\zeta_m(x)$, having the following property:

$$\begin{cases} \operatorname{supp}[\zeta_m] \subset \{x;\ |x - x_0| \le r_0\}, \quad \zeta_m(x) = 1 \quad \text{for} \quad |x - x_0| \le \dfrac{r_0}{2} \\ |\partial^{\alpha}\zeta_m(x)| \le (mcr_0^{-1})^{|\alpha|} \quad \text{for} \quad |\alpha| \le m \end{cases} \tag{24}$$

where c can be considered as an absolute constant (independent of r_0).

Using these two assumptions, we can show that there exists a constant A, depending only on ℓ, such that

$$|\partial^{\alpha}(\zeta_m u)| \le AC\,\alpha!\,(\ell e c r_0^{-1})^m, \quad \text{for} \quad |\alpha| = m, \tag{25}$$

provided that $cr_0^{-1} \ge 2C$.

Hence

$$|\xi^{\alpha}\widehat{\zeta_m u}(\xi)| \le (A' r_0^{\ell})C\,\alpha!\,(\ell e c r_0^{-1})^m \quad \text{for} \quad |\alpha| = m, \tag{26}$$

where A' is also a constant depending only on ℓ. We get

$$|\widehat{\zeta_m u}(\xi)| \le C(Cm)^m/|\xi|^m \quad ({}^{\exists} C), \quad \text{for} \quad |\xi| \ge 1. \tag{27}$$

Conversely if this is true, we see that $u(x)$ is analytic in $|x - x_0| \le \dfrac{r_0}{2}$.

Keeping this in mind, the analytic wave front set is given by: We say $(x_0, \xi^0) \notin WF_A(u)$, if we choose r_0 small and considering a sequence of cut-off functions $\zeta_m(x)$ satisfying (24), then the estimates of the form (27) hold for ξ in a suitable conic neighborhood V_{ξ^0} of ξ^0.

In analogy with the case of $\gamma^{(s)}$ ($s > 1$), we get the similar criterion of the analytic wave front sets. $\{\alpha_n(\xi), \beta_n(x)\}$ is defined as follows:

$$\begin{cases} 1) \quad |\alpha_n^{(\nu)}(\xi)| \le \left(\frac{N}{n} cr_0^{-1}\right)^{|\nu|}, \quad \text{for} \quad |\nu| \le 2N, \quad \text{supp}[\alpha_n] \subset \{\xi;\ |\xi - n\xi^0| \\ \le nr_0\}, \quad \alpha_n(\xi) = 1, \quad \text{for} \quad |\xi - n\xi^0| \le \frac{n}{2} r_0, \\ 2) \quad |\beta_n^{(\nu)}(x)| \le \left(N\, cr_0^{-1}\right)^{|\nu|}, \quad \text{for} \quad |\nu| \le 2N, \quad \text{supp}[\beta_n] \subset \{x;\ |x - x_0| \\ \le r_0\}, \quad \beta_n(x) = 1, \quad \text{for} \quad |x - x_0| \le \frac{r_0}{2} \end{cases}$$

where $N = \frac{1}{2}(ce)^{-2}\,\ell^{-1} r_0^{\,2}$.

Let u be a distribution, put

$$S_n = \sum_{|p+q| \le N} c_{pq}^{\,n} \| \alpha_n^{(p)} \beta_{n\,(q)} u \|, \quad \text{with} \quad c_{pq}^{\,n} = \left(cr_0^{-1}\right)^{|p+q|} n^{-|q|}. \tag{29}$$

Then,

<u>Theorem 5</u>. Suppose $(x_0, \xi^0) \notin WF_A(u)$, then $S_n \le \exp(-\varepsilon n)$, if n is large ($\exists\, \varepsilon > 0$). Here we can put $\varepsilon = c_1 r_0^{\,2}$, where the constant c_1 can be taken as a common constant when r_0 is small, (and as far as $u \in \mathcal{E}'$ runs through an equi-continuous set in $\mathcal{E}'$).

Note. In our arguments, we need to choose r_0 small. In this case, the above character of ε (appearing in S_n) becomes important.

The converse is also true.

Theorem 6. Let $\mathring{S}_n = \sum_{|p| \le N} c_{p0}^n \| \alpha_n^{(p)} \beta_n u \|$. If $\mathring{S}_n \le \exp(-\varepsilon n)$ ($\exists\, \varepsilon > 0$) when n is large, and for some $\{\alpha_n, \beta_n\}$ satisfying (28), then we have $(x_0, \xi^0) \notin WF_A(u)$.

Using these theorems, we can show the analytic-hypoellipticity of parabolic operators. Namely, in (22), we assume $a(x,t;\xi)$ is analytic in x, keeping (21).

The same arguments as in the proof of Proposition 1 can be applied. However in this case fairly careful considerations are required. We do not give here the proof. [8] gives fairly detailed accounts of analytic wave front sets and related topics.

Proposition 2. Suppose $t \mapsto f(x,t)$ is continuous with values in $\mathcal{D}'_x$, and $(x_0, \xi^0) \notin WF_A(f(\cdot,t))$, $\forall t \ge 0$. Then

$$\forall t > 0, \quad (x_0, \xi^0) \notin WF_A(u(\cdot,t)).$$

Appendix

§ A.1. Partition of unity.

In Lecture III § A.3, we explained a partition of unity in $\mathbb{R}^{\ell}_{\xi}$. We consider here the same kind of partition of unity, and therefore use the same notations as there. However, in order to treat Gevrey class functions, we should also use a partition of unity adapted to this class. $\{\phi_k(\rho)\}$ was defined as functions satisfying (III, A.6). Instead of this, we impose the following condition:

$$|\partial^{\alpha+\beta}\phi_k(\rho)| \leq \left(c_0 N_k / R_k\right)^{|\alpha|} \beta!^{1+\varepsilon'} \left(c_0 / R_k\right)^{|\beta|},$$

for $|\alpha| \leq N_k$ $(1 + 2\varepsilon' < s,\ \ \varepsilon' > 0)$. N_k is defined according to problems. Anyway we fix N_k in the form

$$N_k = \theta R_k^{\frac{1}{s}},$$

where $(0 < \theta < 1)$ is a constant.

From this we have, in correspondence with (III, A.7),

$$\partial^{\alpha+\beta}\phi_k(|\xi|) \leq \left(c' N_k / R_k\right)^{|\alpha|} \beta!^{1+\varepsilon'} \left(c' / R_k\right)^{|\beta|},$$

$$|\alpha| \leq N_k,$$

$$\left\{\begin{array}{l} |\partial^{\alpha+\beta}\chi_n(\xi)| \leq \left(c' N' / n\right)^{|\alpha|} \beta!^{1+\varepsilon'} \left(c' / n\right)^{|\beta|} \\ |\partial^{\alpha+\beta}\chi_{n,-}(\xi)| \leq \quad '' \quad '' \\ \quad \text{for } |\alpha| \leq N', \quad N' = \theta\, n^{\frac{1}{s}}. \end{array}\right. \tag{A.1}$$

In what follows, the estimates look complicated. However, they are not so. For instance $p!$ is almost estimated by p^p.

We are mainly concerned with the estimates of commutators. To illustrate such an example, we consider $\zeta(x)\chi_{n,-}(D)$. Let

$$\zeta(x) \in \gamma^{(s)}, \quad |\zeta_{(\nu)}(x)| \leq \nu!^s \, C_0'^{|\nu|}. \tag{A.2}$$

Asymptotic expansion gives

$$\zeta(x)\chi_{n,-} = \sum_{|\nu| \leq N'} (-1)^{|\nu|} \nu!^{-1} \chi_{n,-}^{(\nu)} \zeta_{(\nu)} + s_{N'}. \tag{A.3}$$

From the analysis in Lecture III, Appendix,

$$\| s_{N'} \| \leq c(\ell) \, \nu!^{-1} \left(\frac{N'}{n} c'\right)^{N'} \nu!^s \, C_0'^{|\nu|} \Big|_{|\nu|=N'} \times \left(\#_{(N'+1)}\right),$$

where in general $\#_j$ means the number of $(\nu_1, \cdots, \nu_\ell)$ $(\nu_i \geq 0)$ such that $\nu_1 + \cdots + \nu_\ell$ $(= |\nu|) = j$. $\#_j$ has the estimate $\leq (j+1)^{\ell-1}$.

Hence

$$\| s_{N'} \| \leq \left(\frac{N'^s}{n} c' C_0'\right)^{N'} \cdot c(\ell) \cdot (N' + 2)^{\ell-1}.$$

Now we fix

$$N' = (c' C_0' e)^{-1} \, n^{\frac{1}{s}}. \tag{A.4}$$

Thus

$$\| s_{N'} \| \leq e^{-N'} \cdot c(\ell) \cdot (N' + 2)^{\ell-1}. \tag{A.5}$$

In our argument, it is convenient to introduce

Definition. We say that a sequence a_n (≥ 0) or more generally a sequence of L^2-functions f_n is negligible, if for any large L, a_n or $\|f_n\|$ is estimated by $\exp(-L\, n^{1/s})$ when n is large.

We are mainly concerned with the estimate of the form

$$\exp(-\delta n^{\frac{1}{s}})\times(\text{some polynomial in } n) \quad (\exists\, \delta > 0).$$

We call the second factor trivial factor. Observe that $\#_n$ (the number of ν such that $|\nu| = n$) is a trivial factor, because $\#_n \leq n^{\ell-1}$. Also the number of ν such that $|\nu| \leq n$ is a trivial factor.

Lemma A.1.

$$\sum_{|p+q|\leq N} c^n_{pq} \| \alpha_n{}^{(p)} \beta_{n\,(q)} \chi_{n,\pm} u \| \quad \text{is negligible.}$$

Proof.

$$\alpha_n{}^{(p)} \beta_{n\,(q)} u_{n,-} = \alpha_n{}^{(p)} \beta_{n\,(q)} \chi_{n,-}(D)\, \tilde{u}_n,$$

where $\tilde{u}_n$ is the Fourier inverse image of $\hat{u}_{(n)}(\xi)$, which is equal to $\hat{u}(\xi)$ for $|\xi| \leq n$, and 0 for $|\xi| > n$.

In the right-hand side expression, we commute $\chi_{n,-}(D)$ with $\beta_{n_{(q)}}$ to get

$$\beta_{n_{(q)}} \chi_{n,-} = \sum_{|\nu| \le N_0} (-1)^{|\nu|} \chi_{n,-}^{(\nu)} \beta_{n_{(q+\nu)}} + r_{N_0},$$

where $N_0 = n^{1/s}$.

Since $\alpha_n^{(p)} \chi_{n,-}^{(\nu)} = 0$, $^\forall \nu \ge 0$, $\alpha_n^{(p)} \beta_{n_{(q)}} u_{n,-} = \alpha_n^{(p)} r_{N_0} \tilde{u}_n$. Now we see that $c_{pq}^n \| \alpha_n^{(p)} r_{N_0} \|$ is estimated, except for trivial factor, by

$$e^{-|p+q|} \left(N_0^{1+2\varepsilon'} c' \, cr_0^{-1}/n\right)^{N_0}.$$

Since $N_0^{1+2\varepsilon'} = n^{(1+2\varepsilon')/s}$, and $1 + 2\varepsilon' < s$, the above sequence is evidently <u>negligible</u>. On the other hand,

$$\| \tilde{u}_n \| \le C_\varepsilon \exp\left(\varepsilon n^{\frac{1}{s}}\right) \quad (^\forall \varepsilon > 0).$$

Thus $c_{pq}^n \| \alpha_n^{(p)} \beta_{n_{(q)}} u_{n,-} \|$ is negligible.

Concerning $u_{n,+}$,

$$\| \alpha_n^{(p)} \beta_{n_{(q)}} \chi_{n,+} u \| \le \sum_k{}' \| \alpha_n^{(p)} \beta_{n_{(q)}} \phi_k \tilde{u}_k \|,$$

where $\tilde{u}_k$ is the inverse Fourier image of the function obtained by truncation of $\hat{u}(\xi)$ by the characteristic function of $\{\xi; \frac{1}{3} \cdot 4^k \le |\xi| \le 3 \cdot 4^k\}$. Thus,

$$\| \tilde{u}_k \| \le C_\varepsilon \exp\left(\varepsilon R_k^{\frac{1}{s}}\right) \quad (^\forall \varepsilon > 0).$$

The proof is almost the same as in the case $u_{n,-}$. We omit the proof.

§ A.2. Proof of Theorem 1.

By hypothesis, there exists a $\zeta(x)$, such that (15) holds. We assume

$$|\zeta_{(\nu)}(x)| \le \nu!^{s}\, C_1^{|\nu|} \qquad (^{\exists} C_1 > 0).$$

We assume $C_1 \ge 1$.

Now we consider $c_{pq}^{n} \| \alpha_n^{(p)} \beta_{n_{(q)}} u \|$. We take r_0 to be small, so that $\beta_n \Subset \zeta$ (i.e. on the support of β_n, $\zeta(x) \equiv 1$). Hence $\beta_{n_{(q)}} = \beta_{n_{(q)}} \zeta$. Denoting $u_n = \chi_n u$,

$$c_{pq}^{n}\, \alpha_n^{(p)} \beta_{n_{(q)}} u_n = c_{pq}^{n}\, \alpha_n^{(p)} \beta_{n_{(q)}} \{\zeta u_n\}$$

$$= c_{pq}^{n} \sum_{|\nu| \le N_0} \nu!^{-1} \beta_{n_{(q+\nu)}} \alpha_n^{(p+\nu)} \{\zeta u_n\} + c_{pq}^{n} r_{N_0} \{\zeta u_n\}.$$

We want to show that if we take $N_0 = n^{1/s}$, then $c_{pq}^{n} \| r_{N_0} \|$ is negligible. First, $\| r_{N_0} \|$ is estimated by $\left| \nu!^{-1} \beta_{n_{(q+\nu)}} \alpha_n^{(p+\nu)} \right|_{0, |\nu| = N_0}$, except for a negligible factor. Now

$$c_{pq}^{n} \left| \nu!^{-1} \beta_{n_{(q+\nu)}} \alpha_n^{(p+\nu)} \right|_0 \le e^{-|p+q|} \left\{ |\nu|^{1+2\varepsilon'} \left(c r_0^{-1}\right)^2 / n \right\}^{|\nu|}.$$

Putting $|\nu| = N_0$, $= e^{-|p+q|} \left\{ N_0^{1+2\varepsilon'} \left(c r_0^{-1}\right)^2 / n \right\}^{N_0}$, which is negligible, because $1 + 2\varepsilon' < s$.

Next, we consider one of the above terms under the sum:

$$c^{n}_{pq} \; \nu!^{-1} \beta_{n\,(q+\nu)} \, \alpha_{n}^{(p+\nu)} \left(\zeta u_{n}\right).$$

Now by the result obtained in § A.1 (see (A.4), (A.5)), we see that

$$\| c^{n}_{pq} \; \nu!^{-1} \beta_{n\,(q+\nu)} \, \alpha_{n}^{(p+\nu)} \left(\zeta u_{n,\pm}\right) \|$$

is estimated by the form $e^{-N'} \, C_2 \exp\left(\varepsilon n^{1/s} \right)$ ($\forall \varepsilon > 0$), except for a trivial factor. We showed this only in the case $u_{n,-}$, but in the case $u_{n,+}$ the same argument can be applied considering $\zeta u_{n,+} = \zeta \chi_{n,+} u =$

$= \sum_{k}' \zeta \phi_k \, \tilde{u}_k$ (see the last part of the proof of Lemma A.1.).

Now, we know that, for $\| c^{n}_{pq} \; \nu!^{-1} \beta_{n\,(q+\nu)} \, \alpha_{n}^{(p+\nu)} \left(\zeta u\right) \|$ the hypothesis (15) can be applied directly, which yields that this is estimated by $\exp\left(- \frac{\varepsilon_1}{2} n^{1/s}\right)$, for n large.

Finally, using Lemma A.1,

$$c^{n}_{pq} \| \alpha_{n}^{(p)} \beta_{n\,(q)} u \| \leq c^{n}_{pq} \| \alpha_{n}^{(p)} \beta_{n\,(q)} u_{n} \| + (\text{ negligible })$$

$$\leq \exp\left(- \frac{\varepsilon}{2} n^{1/s}\right) \times (\text{ trivial factor }) + e^{-N'/2} + (\text{ negligible }).$$

(Recall that N' is defined by (A.4) $= \left(c' C_1 e\right)^{-1} n^{1/s}$.) Thus the proof of Theorem 1 is complete.

§ A.3. Proof of (18).

We show that

$$\| \alpha_n \zeta u \| \leq \tilde{S}_n + (\text{ negligible }), \tag{A.6}$$

$u = u_n + u_{n,-} + u_{n,+}$ $(= \chi_n u + \chi_{n,-} u + \chi_{n,+} u)$, and we know that $\| \alpha_n \zeta u_{n,\pm}$ is negligible, because this is essentially the same as in Lemma A.1.

Next,

$$\| \alpha_n \zeta u_n \| = \| \alpha_n \zeta (\beta_n u_n) \|$$

$$\leq \sum_{|\nu| \leq N} \nu!^{-1} |\zeta_{(\nu)}|_0 \| \alpha_n^{(\nu)} \beta_n u_n \| + \| r_N \| \, \| \beta_n u_n \| .$$

First, $\| r_N \|$ is negligible. Next, in order to compare $\sum_{|\nu| \leq N} \cdots$, with $\tilde{S}_n$, we consider

$$\nu!^{-1} |\zeta_{(\nu)}|_0 \, n^{-(1-\frac{1}{s})|\nu|} \leq \nu!^{\varepsilon'} C_1^{|\nu|} n^{-(1-\frac{1}{s})|\nu|}$$

$$\leq \left(|\nu|^{\varepsilon'} C_1 n^{-(1-\frac{1}{s})} \right)^{|\nu|} \leq 1.$$

Thus we proved (A.6).

Note. In almost all statements, we said " if n is large ", and did not make precisely from what n that holds. Take (18) as an example. If you look at the above proof, you realize that this analysis is closely connected to the notion of negligible. As an example of the negligible quantity if we look at $\left(\frac{N^{1+2\varepsilon'}}{n} (c r_0^{-1})^2 \right)^N$ $(N = (ce)^{-1} r_0,\ 1 + 2\varepsilon' < s)$, we see that this quantity is not uniformly small with respect to r_0 (when n tends to ∞). It is important to notice that the terminology " negligible " means that <u>once r_0 is fixed</u> (however small it may be) and n tends to ∞, this quantity becomes negligible (i.e. estimated by $\exp(-L n^{1/s})$) , so that (18) means that ε_0' is a constant independent of r_0, however its validity is only for $n \geq n_0(r_0)$ and when r_0 becomes smaller and smaller, we should take $n_0(r_0)$ to be larger and larger.

§ A.4. Pseudo-local property in $\gamma^{(s)}$.

Let $a(x, \xi)$ be a Gevrey-pseudoanalytic symbol of class s. To be precise,

$$|a_{(\beta)}^{(\alpha)}(x, \xi)| \leq A\, \alpha!\, \beta!^{s}\, C_0^{|\alpha+\beta|} |\xi|^{m-|\alpha|} \text{ for } |\alpha| \leq R^{-1}|\xi|^{1/s}, \quad (A.7)$$

where R is a suitable large number.

As an application of Theorems 1 and 2, we prove the following theorem which is known, however its proof serves as the core part of the proof of Fundamental Theorem in Gevrey class.

<u>Theorem A.1</u>. Let $u \in (\gamma^{(s)})'$ with compact support. If $(x_0, \xi^0) \notin WF_s(u)$, then $(x_0, \xi^0) \notin WF_s(a(x,D)u)$.

To prove this, we show

Lemma A.2.

$$\sum_{|p+q|\le N} c_{pq}^{n} \| \alpha_n^{(p)} \beta_{n(q)} a(x,D) u_{n,\pm} \| \le \exp(-\varepsilon_1 n^{1/s}) \quad (\exists \varepsilon_1 > 0),$$

if n is large, where ε_1 can be taken to be independent of r_0 when r_0 is small.

Observe that this lemma is independent of the assumption of Theorem A.1. The principle of the proof is almost the same as that of Lemma A.1.

Proof. As in the proof of Lemma A.1, we prove this lemma for $u_{n,-}$.

$$\alpha_n^{(p)} \beta_{n(q)} a(x,D) u_{n,-} = \alpha_n^{(p)} \beta_{n(q)} a \chi_{n,-} \tilde{u}_n,$$

where $\tilde{u}_n$ is defined in the proof of Lemma A.1. We commute a with $\chi_{n,-}$. Before doing this, let $\chi(\xi) \in C_0^\infty$ $(0 \le \chi(\xi) \le 1)$, $= 1$ for $|\xi| \le 1$, and $= 0$ for $|\xi| \ge 2$.

$$a\chi_{n,-} = a\chi(\tfrac{D}{n})\chi_{n,-}.$$

Denote $a_n = a\chi(\frac{D}{n})$. Then

$$a\chi_{n,-} = a_n \chi_{n,-} = \sum_{|\mu|\le N'} (-1)^{|\mu|} \mu!^{-1} \chi_{n,-}^{(\mu)} a_{n(\mu)} + s_{N'}.$$

We see easily that $\| s_{N'} \|$ is estimated, except for trivial factor, by

$$\left(\frac{N'}{n} c'\right)^{N'} \left(N'^{s-1} C_0\right)^{N'} = \left(\frac{N'^s}{n} c' C_0\right)^{N'}.$$

Now we fix, in (A.1), $N' = (c'C_0e)^{-1} n^{1/s}$. We can assume $c'C_0e > R$ (see (A.7)), by taking C_0 to be large if necessary. Hence,

$$\| s_{N'} \| \text{ is estimated by } e^{-N'} \ (= \exp[-(c'C_0e)^{-1} n^{1/s}]) \\ \text{except for a trivial factor if } n \text{ is large.} \tag{A.8}$$

Now

$$c^n_{pq}\, \alpha_n{}^{(p)} \beta_{n_{(q)}}\, a\chi_{n,-}\tilde{u}_n = \sum_{|\mu| \le N'} (-1)^{|\mu|} \mu!^{-1}\, c^n_{pq}\, \alpha_n{}^{(p)} \beta_{n_{(q)}}\, \chi_{n,-}{}^{(\mu)} a_{n_{(\mu)}} \tilde{u}_n \\ + c^n_{pq}\, \alpha_n{}^{(p)} \beta_{n_{(q)}}\, s_{N'}\, \tilde{u}_n. \tag{A.9}$$

Show that each term $\mu!^{-1}\, c^n_{pq}\, \alpha_n{}^{(p)} \beta_{n_{(q)}}\, \chi_{n,-}{}^{(\mu)} a_{n_{(\mu)}} \tilde{u}_n$ is negligible.

Denoting $N_0 = n^{1/s}$,

$$\beta_{n_{(q)}}\, \chi_{n,-}{}^{(\mu)} = \sum_{|\nu| \le N_0} (-1)^{|\nu|} \nu!^{-1} \chi_{n,-}{}^{(\mu+\nu)} \beta_{n_{(q+\nu)}} + r_{N_0}.$$

Observe that $\alpha_n{}^{(p)} \chi_{n,-}{}^{(\mu+\nu)} = 0$. Thus, the above term is expressed by

$$\mu!^{-1}\, c^n_{pq}\, \alpha_n{}^{(p)} r_{N_0}\, a_{n_{(\mu)}} \tilde{u}_n.$$

Now $\| r_{N_0} \|$ is estimated, except for a trivial factor, by

$$\nu!^{-1} \cdot \left(\frac{N'}{n} c'\right)^{|\mu|} \nu!^{1+\varepsilon'} \left(\frac{c'}{n}\right)^{|\nu|} (Ncr_0^{-1})^{|q|} \nu!^{1+\varepsilon'} (cr_0^{-1})^{|\nu|} \Big|_{|\nu| = N_0}$$

$$\le \left(\frac{N'}{n}\right)^{|\mu|} \left(N_0^{1+2\varepsilon'} c'cr_0^{-1}/n \right)^{N_0} (cr_0^{-1})^{|q|}.$$

Hence, $\left\| u!^{-1}\, c^{n}_{pq}\, \alpha_n^{(p)} \tau_{N_0}\, a_{n_{(\mu)}}\, \tilde{u}_n \right\|$ is estimated, except for a trivial factor, by

$$\underbrace{\mu!^{s-1} C_0^{|\mu|} (2n)^m}_{\mu!^{-1} a_{n_{(\mu)}}} e^{-|p+q|} \left(\frac{N'}{n}\right)^{|\mu|} \left(N_0^{1+2\varepsilon'} c' c r_0^{-1}/n\right)^{N_0} \| \tilde{u}_n \| .$$

Since $|\mu| \le N'$, and $(2n)^m$ is a trivial factor, this is again estimated, except for a trivial factor, by

$$e^{-|p+q|} \left(\frac{N'^s}{n} C_0 c'\right)^{|\mu|} \left(N_0^{1+2\varepsilon'} c' c r_0^{-1}/n\right)^{N_0} \| \tilde{u}_n \| ,$$

where we assume $c' \ge 1$.

Taking account of (A.8), $\| c^n_{pq}\, \alpha_n^{(p)} \beta_{n_{(q)}}\, a \chi_{n,-} \tilde{u}_n \|$ is estimated, except for a trivial factor, by

$$e^{-|p+q|} \left(\sum_{|\mu| \le N'} \left(\frac{N'^s}{n} C_0 c'\right)^{|\mu|} \right) \left(N_0^{1+2\varepsilon'} c' c r_0^{-1}/n\right)^{N_0} \| \tilde{u}_n \| + e^{-|p+q|} e^{-N'} \| \tilde{u}_n \|$$

Thus, observing $\sum_{|\mu| < \infty} e^{-|\mu|} < +\infty$, this is estimated by (except for trivial factor)

$$e^{-|p+q|} \| \tilde{u}_n \| \left[\left(N_0^{1+2\varepsilon'} c' c r_0^{-1}/n\right)^{N_0} + e^{-N'} \right].$$

Thus,

$$\sum_{|p+q| \le N} c^n_{pq} \| \alpha_n^{(p)} \beta_{n_{(q)}}\, a \chi_{n,-} \tilde{u}_n \|$$

is estimated by (except for a trivial factor)

$$e^{-N'}\|\tilde{u}_n\| + \left(N_0^{1+2\varepsilon'}c'cr_0^{-1}/n\right)^{N_0}\|\tilde{u}_n\| . \tag{A.10}$$

The second term is evidently negligible, which proves Lemma A.2. □

§ A.5. Proof of Theorem A.1.

By Theorem 1, the assumption implies

$$S_n(u) = \sum_{|p+q|\le N} c_{pq}^n \|\alpha_n^{(p)}\beta_{n_{(q)}} u\| \le \exp\left(-\varepsilon_1 n^{1/s}\right),$$

where ε_1 (> 0) is independent of r_0 when r_0 is small. Taking account of Theorem 2, it is enough to show that

$$S_n(au) = \sum_{|p+q|\le N} c_{pq}^n \|\alpha_n^{(p)}\beta_{n_{(q)}} au\| \le \exp\left(-\varepsilon_1 n^{1/s}\right) \quad (\exists\, \varepsilon_1 > 0) \tag{A.11}$$

when r_0 is small, and n tends to ∞ .

This is reduced to consider the commutators $[\alpha_n^{(p)}\beta_{n_{(q)}}, a(x,D)]$, $(|p+q| \le N)$. We decompose

$$u = u_n + u_{n,-} + u_{n,+} \quad (\equiv \chi_n u + \chi_{n,-}u + \chi_{n,+}u).$$

Now Lemma A.2 shows that $\sum_{|p+q|\le N} c_{pq}^{n} \| \alpha_n^{(p)} \beta_{n_{(q)}} a u_{n,\pm} \|$ is estimated by $\exp\left(-\varepsilon_2 n^{1/s}\right)$ ($\exists\, \varepsilon_2 > 0$), where ε_2 is also independent of r_0. It is also almost evident that

$$\mathring{S}_n = \sum_{|p+q|\le N} c_{pq}^{n} \| a\alpha_n^{(p)} \beta_{n_{(q)}} u_{n,\pm} \| \quad \text{is negligible.} \tag{A.12}$$

In fact, we consider the commutators $[\beta_{n_{(q)}}, \chi_{n,-}]$, $[\beta_{n_{(q)}}, \phi_k]$. In this case, we can easily see (A.11) is true.

We consider therefore, in (A.11), u_n instead of u. We denote $a\chi_n = a_n$, it is easy to see that

$$\left\| a_{n_{(\mu)}}^{(\nu)} \right\| \le A\, \mu!^{s}\, \nu!^{1+\varepsilon'}\, C_0^{|\mu+\nu|}\, n^{m} \left(\frac{n}{2}\right)^{-|\nu|}. \tag{A.13}$$

We denote by $\tilde{u}_n$ the Fourier inverse image of $\chi_{(n)}(\xi)\hat{u}(\xi)$, where $\chi_{(n)}$ is the characteristic function of $\operatorname{supp}[\chi_n]$. We have

$$a\, u_n = a\chi_n u = a\chi_n \tilde{u}_n = a_n \tilde{u}_n.$$

Now (see III, § A.6),

$$\begin{aligned} c_{pq}^{n}\, \alpha_n^{(p)} \beta_{n_{(q)}} a u_n &= c_{pq}^{n}\, \alpha_n^{(p)} \beta_{n_{(q)}} a_n \tilde{u}_n \\ &= c_{pq}^{n} \sum_{|\mu+\nu|\, N-(p+q)} \mu!^{-1} (-1)^{|\nu|} \nu!^{-1} a_{n_{(\mu)}}^{(\nu)} \alpha_n^{(p+\mu)} \beta_{n_{(q+\nu)}} \tilde{u}_n \\ &\qquad + c_{pq}^{n}\, s_{N,p,q}\, \tilde{u}_n. \end{aligned} \tag{A.14}$$

We assume hereafter $C_0 \ge c > 1$. Observe that in (A.7) we assume $|\nu| \le R^{-1} n^{1/s}$. This is realized because $N = (ce)^{-1} r_0 n^{1/s}$, and r_0 is small.

(1st step) Estimate of $c^n_{pq} \| s_{N,p,q} \|$.

This is estimated except for a trivial factor by

$$c^n_{pq}\, \mu!^{-1} \nu!^{-1} \left| a_n{}^{(\nu)}{}_{(\mu)} \right|_0 \left| \alpha_n{}^{(p+\mu)} \right|_0 \left| \beta_n{}_{(q+\nu)} \right|_0 ,$$

by putting $|p + q + \mu + \nu| = N$. This is estimated, by using (A.3) (except for a trivial factor),

$$e^{-|p+q|} \mu!^{s-1} \nu!^{\varepsilon'} \, C_0{}^{|\mu+\nu|} \left(\frac{2}{n}\right)^{|\nu|} \left(\frac{N}{n} cr_0{}^{-1}\right)^{|\mu|} \left(Ncr_0{}^{-1}\right)^{|\nu|}$$

$$\leq e^{-|p+q|} \left(\frac{N}{n} |\mu|^{s-1} \, C_0 cr_0{}^{-1}\right)^{|\mu|} \left(\frac{1}{n} N^{1+\varepsilon'} \, 2C_0 cr_0{}^{-1}\right)^{|\nu|}$$

$$\leq e^{-|p+q|} \left(\frac{1}{n} N^s \, C_0 cr_0{}^{-1}\right)^{|\mu|} \left(\frac{1}{n} N^{1+\varepsilon'} \, 2C_0 cr_0{}^{-1}\right)^{|\nu|}$$

$$\leq e^{-|p+q+\mu+\nu|} = e^{-N}, \quad \underline{\text{if } r_0 \text{ is small}}.$$

In fact, $\frac{1}{n} N^s C_0 cr_0{}^{-1} \leq C_0 r_0{}^{s-1} e^{-1}$, hence we assume

$$r_0 \leq \left(C_0{}^{-1}\right)^{\frac{1}{s-1}} \tag{A.15}$$

Concerning the factor $\left(\frac{1}{n} N^{1+\varepsilon'} \, 2C_0 cr_0{}^{-1}\right)^{|\nu|} \leq e^{-|\nu|}$ is realized if (A.15) is assumed.

Summing up.

$c^n_{pq} \| s_{N,p,q} \| \leq e^{-N} \times$ (some fixed polynomial in n), if (A.15) is assumed.

To be precise, this is true for $n \geq n_0$, which is independent of (p, q).

(2nd step) Look at (A.14).

$$c_{pq}^{n}\| \alpha_n^{(p)} \beta_{n(q)} au_n \| \le c_{pq}^{n} \sum_{|\mu+\nu|\le N-(p+q)} \| \mu!^{-1}\nu!^{-1} a_{n(\mu)}^{(\nu)} \| \, \| \alpha_n^{(p+\mu)} \beta_{n(q+\nu)} \tilde{u}_n \|$$

$$+ c_{pq}^{n} \| s_{N,p,q} \| \, \| \tilde{u}_n \| .$$

Now,

$$\| \mu!^{-1}\nu!^{-1} a_{n(\mu)}^{(\nu)} \| \, c_{pq}^{n} \Big/ c_{p+\mu,q+\nu}^{n} = \| \cdots \| \Big/ n^{(1-\frac{1}{s})|\nu| - \frac{1}{s}|\mu|} .$$

Here we use (A.13). This is estimated by

$$A\, n^{m} \left(C_0 |\mu|^{s-1} \Big/ n^{1-\frac{1}{s}} \right)^{|\mu|} \left(2C_0 |\nu|^{\varepsilon'} \Big/ n^{1-\frac{1}{s}} \right)^{|\nu|} . \tag{A.16}$$

Now

$$C_0 |\mu|^{s-1} \Big/ n^{1-\frac{1}{s}} \le C_0 N^{s-1} \Big/ n^{1-\frac{1}{s}} = C_0 (ce)^{-(s-1)} r_0^{\,s-1} .$$

We assumed (A.15). Then this is less than $(ce)^{-(s-1)} < 1$. It is the same with the second factor.

Summing up.

$$c_{pq}^{n} \| \alpha_n^{(p)} \beta_{n(q)} au_n \| \le A\, n^{m} \sum_{|\mu+\nu|\le N-(p+q)} c_{p+\mu,q+\nu}^{n} \| \alpha_n^{(p+\mu)} \beta_{n(q+\nu)} \tilde{u}_n \|$$

$$+ c_{pq}^{n} \| s_{N,p,q} \| \, \| \tilde{u}_n \| \tag{A.17}$$

where $\| \tilde{u}_n \| \le C_\varepsilon \exp(\varepsilon n^{1/s})$ ($\forall\, \varepsilon > 0$).

(3rd step) It remains to consider the estimates of the difference when we replace, in (A.17), $\| \alpha_n^{(p+\mu)} \beta_{n_{(q+\nu)}} \tilde{u}_n \|$ by $\| \alpha_n^{(p+\mu)} \beta_{n_{(q+\nu)}} u \|$.

First, let us recall the definition of $\tilde{u}_n$ ($= \chi_{(n)} u$), and $u_n = \chi_n u$. Since on $\mathrm{supp}[\alpha_n]$, $\chi_n(\xi) = 1$, and $\chi_{(n)}(\xi) = 1$, $\mathrm{supp}[\alpha_n] \cap \mathrm{supp}[\chi_n - \chi_{(n)}] = \phi$. Thus, we consider $\alpha_n^{(p+\mu)} \beta_{n_{(q+\nu)}} (\chi_n - \chi_{(n)}) u$. By commuting $\alpha_n^{(p+\mu)}$ with $\beta_{n_{(q+\nu)}}$, we see that $c^n_{p+\mu,q+\nu} \| \alpha_n^{(p+\mu)} \beta_{n_{(q+\nu)}} (\chi_n - \chi_{(n)}) u \|$ is estimated by

$$e^{-|p+q+\mu+\nu|} \left(\frac{1}{n} N_0^{1+2\varepsilon'} (cr_0^{-1})^2 \right)^{N_0} \times \| u_n - \tilde{u}_n \| \qquad (N_0 = n^{1/s})$$

(except for a trivial factor). Thus, we arrived at the conclusion that, in (A.17), (since $\| u_n - \tilde{u}_n \| \le C_\varepsilon \exp(\varepsilon n^{1/s})$, $\forall \varepsilon > 0$), in the sum $\sum \cdots$, when we replace $\tilde{u}_n$ by u_n, the difference is negligible. Now, (A.12) says that even when we replace u_n by u the difference is also negligible.

Thus, from (A.17), we get

$$S_n(au) \le \exp(-\varepsilon_2 n^{1/s}) + S_n(u) \times (\text{some polynomial}) + e^{-N/2}, \tag{A.18}$$

when $n \to \infty$, if r_0 is small (see (A.15)). Observe that ε_2 is independent of r_0. Thus, taking account of the assumption of $S_n(u)$, we can say that if r_0 is small,

$$S_n(au) \le 2\, e^{-N/2},$$

if n is large. Since $n = (ce)^{-1} r_0 n^{1/s}$, this proves Theorem.

References

[1] A. Friedman, Regularity of fundamental solutions of hyperbolic equations Arch. Math. Mech., 11(1962), 62-96.

[2] L. Hörmander, Uniqueness theorems and wave front sets for solutions of linear differential equations with analytic coefficients, Comm. Pure Appl. Math., 24(1971), 671-704.

[3] K. Kajitani, Leray-Volevich's system and Gevrey class, J. Math. Kyoto Univ., 21(1981), 547-74.

[4] H. Komatsu, Ultradistributions, I, Structure theorems and a Characterization, J. Fac. Sci. Univ. Tokyo, Sec. IA, 20(1973), 25-105.

[5] S. Mizohata, Propagation des singularités au sens de Gevrey pour les opérateures différentîels à multiplicité constante, Séminaire sur les équations aux dérivées partielles de J. Vaillant(1982-83), 106-33, Travaux en cours, Hermann, 1984.

[6] L. Nirenberg, Lectures on linear partial differential equations, Regiona Conference Series in Mathematics, 17(1973), Amer. Math. Soc.

[7] K. Taniguchi, Fourier integral operators in Gevrey class on $\mathbb{R}^n$ and the fundamental solution for a hyperbolic operator, to appear in Pub. Res. Inst. Math. Sci. Kyoto Univ.

[8] S. Mizohata, On analytic regularities, to appear in Séminaire sur les équations aux dérivées partielles de J. Vaillant(1983-84), Hermann.

Lecture VI Micro-local Analysis in Gevrey Class (II).

§ 1. Preliminaries.

In the last lecture, we explained our view-point. Prof. Chen Shuxing pointed out that the essential point is the pseudo-local property of the pseudo-differential operators. This is ture. However, we should add some comments to this. In the C^∞-micro-local analysis, we can treat almost all the problems from this view-point. In other words, the qualitative statement will be enough. Contrary to this, in Gevrey class micro-local analysis, we are forced to take another view-point: quantitative view-point. Then, in general, the argument becomes heavy and tedious. Our first purpose is to prove Fundamental Theorem in Gevrey class announced in Lecture V. We repeat this. Let

$$L(u) \equiv \left(\partial_t + i\lambda(x,t;D) + c(x,t;D)\right) u(x,t) = f(x,t), \qquad (1)$$

where we assume for $\lambda(x,t;\xi)$, $c(x,t;\xi)$ the same conditions as (IV,9) and also (V, A.7) for λ and c. Let $(x(t), \xi(t))$ be the bicharacteristic strip of L (see Figure). Then

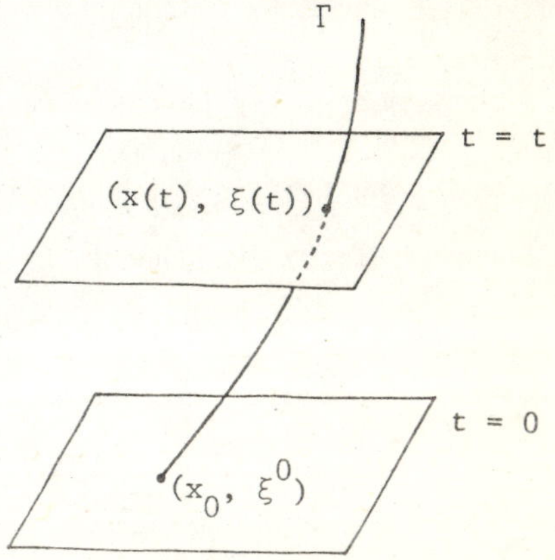

Theorem 1. If $(x_0, \xi^0) \underset{(\in)}{\notin} WF_s(u(\cdot,0))$ and $(x(t), \xi(t)) \notin WF_s(f(\cdot,t))$, then it follows

$$(x(t), \xi(t)) \underset{(\in)}{\notin} WF_s(u(\cdot,t)), \quad {}^\forall t \geq 0.$$

In order to prove Theorem 1, we use some micro-localizer along the bicharacteristic strip Γ. In the C^∞-case we explained this in Lecture V (last part of § 1). Instead of (V; 9), in actual case, we take a series of cut-off functions $\{\alpha_n(\xi,t),\ \beta_n(x,t)\}$, depending smoothly on t, as a solution

$$\{\partial_t + \lambda_\xi(t)\partial_x - n\,\lambda_x(t)\partial_\xi\}\{\alpha_n(\xi,t)\beta_n(x,t)\} = 0, \qquad t \geq 0.$$

$$\text{with} \quad \begin{cases} \alpha_n(\xi,0) = \alpha_n(\xi) \\ \beta_n(x,0) = \beta_n(x) \end{cases} \tag{2}$$

where

$$\begin{cases} \lambda_\xi(t) = \lambda_\xi\{x(t),t;\ \xi(t)\} \\ \lambda_x(t) = \lambda_x\{x(t),t;\ \xi(t)\}. \end{cases} \tag{3}$$

Observe that $\{\alpha_n(\xi),\ \beta_n(x)\}$ is defined by (V; 11), (V; 12) and (V; 13). More concretely, we define $\alpha_n(\xi,t)$, $\beta_n(x,t)$ by

$$\begin{cases} \partial_t\alpha_n - n\,\lambda_x(t)\partial_\xi\alpha_n = 0 \\ \partial_t\beta_n + \lambda_\xi(t)\partial_x\beta_n = 0 \end{cases} \tag{4}$$

with initial data (2).

We see easily that the assumptions on $\{\alpha_n(\xi),\ \beta_n(x)\}$, (V; 11) ~ (V; 13) remain true for $\{\alpha_n(\xi,t),\ \beta_n(x,t)\}$. We take $|\xi^0| = 1$, in that case, $|\xi(t)|$ is not strictly 1, however if we limit to t fairly small, $|\xi(t)|$ remains close to 1, so that in our arguments there is no serious modification.

In view of Theorem 1 and Theorem 2 in Lecture V, we carry out the proof in the following way.

Put

$$S_n(t) = \sum_{|p+q| \le N} c_{pq}^n \| \alpha_n^{(p)} \beta_{n(q)} u(\cdot,t) \| . \tag{5}$$

First, we suppose that $(x_0, \xi^0) \notin WF_s(u(\cdot,0))$. Then, from Theorem 1 in Lecture V, we get

$$S_n(0) \le \exp\left(-\varepsilon_0 n^{1/s}\right) \tag{6}$$

for n large.

In this case, ε_0 (> 0) is a constant which can be regarded as the one independent of r_0, if r_0 is taken small. Recall that r_0 is the size of cut-off functions. Our aim is to show the same type of estimate for $S_n(t)$ holds (see Theorem 2 in Lecture V). More precisely,

$$S_n(t) \le \exp\left(-\varepsilon_1 n^{1/s}\right), \quad \text{if} \quad 0 \le t \le t_0 \quad (\exists t_0 > 0).$$

The arguments which follow show that this is possible if t_0 is taken small, and t_0 can be considered as independent of the initial data, more concretely, independent of the positive number ε_0 which appears in (6). In the proof, the same kind of arguments as in the proof of Theorem A.1 in Lecture V can be applied. In this sense, we could say that, from our view-point, the property of the propagation of regularity (or singularity) along bicharacteristic strips is an immediate consequence of the pseudo-local property (in Gevrey class) of pseudo-differential operators of Gevrey class.

§ 2. Proof of Theorem 1.

Put

$$i\lambda + c = a.$$

We operate $\alpha_n^{(p)}\beta_{n(q)}$ to (1) $(|p+q| \leq N)$, then

$$L\left(\alpha_n^{(p)}\beta_{n(q)}u\right) = \left(\partial_t\alpha_n^{(p)}\right)\beta_{n(q)}u + \alpha_n^{(p)}\left(\partial_t\beta_{n(q)}\right)u + [\alpha_n^{(p)}\beta_{n(q)}, a]\,u$$

$$+ \alpha_n^{(p)}\beta_{n(q)}f.$$

We denote the right-hand side by $g_{n,p,q}$. Taking account of (4), and taking the first terms of the asymptotic expression of commutator, we get

$$\begin{aligned} g_{n,p,q} = & \sum_j \left(n\lambda_{x_j}(t) - \lambda_{x_j}(x,t;D)\right)\left(\alpha_n^{(p+e_j)}\beta_{n(q)}u\right) \\ & - i\sum_j \left(\lambda_{\xi_j}(t) - \lambda_{\xi_j}(x,t;D)\right)\left(\alpha_n^{(p)}\beta_{n(q+e_j)}u\right) \\ & + \sum_j c_{(e_j)}\left(\alpha_n^{(p+e_j)}\beta_{n(q)}u\right) - \sum_j c^{(e_j)}\left(\alpha_n^{(p)}\beta_{n(q+e_j)}u\right) \\ & + C_{n,p,q}(u) + \alpha_n^{(p)}\beta_{n(q)}f. \end{aligned} \tag{7}$$

We divide the consideration for $g_{n,p,q}$ in several steps.

1) By Lemma A.2 in Lecture V, and (V; A.12), in the expression of $g_{n,p,q}$, if we replace u by u_n ($= \chi_n(D)u$), and denote this by $\tilde{g}_{n,p,q}$, the difference $g_{n,p,q} - \tilde{g}_{n,p,q} = h_{n,p,q}$ has the following estimate

$$\sum_{|p+q|\le N} c^n_{pq} \| h_{n,p,q} \| \le \exp\left(-\varepsilon_2 n^{1/s}\right) \quad (\exists \varepsilon_2 > 0) \tag{8}$$

where ε_2 is independent of r_0 when r_0 is small.

2) We denote

$$\mathring{S}_n(t) = \sum_{|p+q|\le N} c^n_{pq} \| \alpha_n^{(p)} \beta_{n(q)} u_n \| .$$

Then,

$$|S_n(t) - \mathring{S}_n(t)| \le \exp\left(-\varepsilon_3 n^{1/s}\right) \quad (\exists \varepsilon_3 > 0) \tag{9}$$

where ε_3 is also independent of r_0 when r_0 is small(Lemma A.2 in Lecture V).

3) When n is large, we have

$$\begin{aligned} &\Big\| \sum_j \left(n\lambda_{x_j}(t) - \lambda_{x_j}(x,t;D)\right)\alpha_n^{(p+e_j)} \beta_{n(q)} u_n \Big\| \\ &\qquad\qquad \le 2nAr_0 \sum_j \| \alpha_n^{(p+e_j)} \beta_{n(q)} u_n \| + \varepsilon_{n,p,q}, \\ &\Big\| \sum_j \left(\lambda_{\xi_j}(t) - \lambda_{\xi_j}(x,t;D)\right)\alpha_n^{(p)} \beta_{n(q+e_j)} u_n \Big\| \\ &\qquad\qquad \le 2Ar_0 \sum_j \| \alpha_n^{(p)} \beta_{n(q+e_j)} u_n \| + \varepsilon'_{n,p,q} \end{aligned} \tag{10}$$

where $\sum_{|p+q|\le N} c^n_{pq}\, \varepsilon_{n,p,q}$, $\sum_{|p+q|\le N} c^n_{pq}\, \varepsilon'_{n,p,q}$ are both negligible.

Concerning the term " negligible ", its definition is given in § A.1 in Lecture V. The constant A denotes the bound of $|\lambda_{x_j}(x,t;\xi)|$, $|\lambda_{\xi_j}(x,t;\xi)|$ $(1 \le j \le \ell)$, over $x \in \mathbb{R}^\ell$, $\frac{1}{2} \le |\xi| \le 1$, and $t \in [0, 1]$.

Roughly speaking, the above estimates can be derived by applying sharp Gårding inequality. Since this estimate is crucial to out argument, we give its proof at the end of this section.

4)

$$\left\| \sum_j c_{(e_j)} \left(\alpha_n^{(p+e_j)} \beta_{n(q)} u_n\right) - \sum_j c^{(e_j)} \left(\alpha_n^{(p)} \beta_{n(q+e_j)} u_n\right) \right\| \tag{11}$$

$$\le \text{const.} \left\{ n^\rho \sum_j \| \alpha_n^{(p+e_j)} \beta_{n(q)} u\| + n^{\rho-1} \sum_j \| \alpha_n^{(p)} \beta_{n(q+e_j)} u_n \| \right\},$$

where const. is independent of (n,p,q,r_0).

5)

$$\sum_{|p+q| \le N} c_{pq}^n \| C_{n,p,q}(u_n) \| \le \text{const.} \frac{n^{1/s}}{n^{1-1/s}} \overset{\circ}{S}_n(t) + e^{-N/2} + (\text{negligible}). \tag{12}$$

This estimate is also crucial to our argument, especially exponential term $e^{-N/2}$ is important. Its proof is the copy of the one of Theorem A.1 in Lecture V. However we give the proof at the end of this section.

6)

$$\sum_{|p+q| \le N} c_{pq}^n \| \alpha_n^{(p)} \beta_{n(q)} f(\cdot,t) \| \le \exp\left(-\varepsilon' n^{1/s}\right) \quad (\exists \varepsilon' > 0) \tag{13}$$

where ε' is independent of r_0 when r_0 is small. □

Now we use fundamental fact:

Energy inequality.

For $L(\alpha_n^{(p)}\beta_{n_{(q)}}u) = g_{n,p,q}$, we have

$$\frac{d}{dt}\|\alpha_n^{(p)}\beta_{n_{(q)}}u\| \leq \text{const. } n^{\rho}\|\alpha_n^{(p)}\beta_{n_{(q)}}u\| + \|g_{n,p,q}\|. \tag{14}$$

First, from this inequality, we have

$$\frac{d}{dt}S_n(t) \leq \text{const. } n^{\rho}S_n(t) + \sum_{|p+q|\leq N} c_{pq}^{n}\|g_{n,p,q}\|$$

$$\leq \text{const. } n^{\rho}S_n(t) + \sum_{|p+q|\leq N} c_{pq}^{n}\|\tilde{g}_{n,p,q}\| + \sum_{|p+q|\leq N} c_{pq}^{n}\|h_{n,p,q}\|.$$

From the above estimate (10) ~ (13), we get

$$\frac{d}{dt}S_n(t) \leq (2\ell A)r_0\, n^{1/s}\,\mathring{S}_n(t) + \text{const. } n^{\rho-1+1/s}\,\mathring{S}_n(t)$$

$$+ \text{const.}\frac{n^{1/s}}{n^{1-1/s}}\,\mathring{S}_n(t) + e^{-N/2} + \exp\left(-\varepsilon' n^{1/s}\right) + (\text{negligible}).$$

Using (9), and denoting $\max(\rho - 1, \frac{1}{s} - 1) = -\delta \quad (\delta > 0)$, we get

$$\frac{d}{dt}S_n(t) \leq \left((2\ell A)r_0 + \text{const. } n^{-\delta}\right) n^{1/s}S_n(t) + e^{-N/2} + \exp\left(-\varepsilon' n^{1/s}\right)$$

$$+ (\text{negligible}).$$

Hence we have, denoting

Hence we have, denoting

$$2\ell A = A',$$

$$S_n(t) \leq \exp\left[\left(A'r_0 + \text{const. } n^{-\delta}\right)n^{1/s}t\right]\Big\{S_n(0)$$

$$+ t\left(e^{-N/2} + \exp(-\varepsilon' n^{1/s}) + (\text{negligible})\right)\Big\}.$$

Now, let us recall

$$N = (ce)^{-1} r_0 n^{1/s}.$$

First, $S_n(0)$ has the favorable estimate (6). Recall that ε_0 is independent of r_0 when r_0 becomes small. We fix r_0 (> 0) in such a way that

$$A' r_0 < \frac{1}{2}\varepsilon_0.$$

Recall that we have already argued assuming r_0 is small. Next we take t_0 (≤ 1) in such a way that

$$A' t_0 \leq \frac{1}{4}(ce)^{-1}.$$

Namely

$$t_0 = \min\left\{(4A'ce)^{-1}, 1\right\}. \tag{15}$$

Then, we see clearly, for $t \in [0, t_0]$, the following type estimate holds

$$S_n(t) \leq \exp\left(-\varepsilon_1 n^{1/s}\right)$$

if n is large, which proves Theorem 1. Observe that t_0 defined by (11) depends essentially on A (= bound of first derivatives of $\lambda(x,t;\xi)$). Thus, we can see that the conclusion holds for any t, by applying the above result a finite times. Next, when we suppose $(x_0, \xi^0) \in WF_s(u(\cdot,0))$, and suppose for some $t > 0$, $(x(t), \xi(t)) \notin WF_s(u(\cdot,t))$. Then, we can argue in the same way as above for negative direction of t, and would obtain $(x_0, \xi^0) \notin WF_s(u(\cdot,0))$, which contradicts the assumption. □

Proof of (10). We define the mapping $x \mapsto \tilde{x}(x,t) \in C^\infty$ in the following way: $\tilde{x} = x$ for $|x - x(t)| \le r_0$, $|\tilde{x} - x(t)| \le 2r_0$, and $\tilde{x} = x(t)$ for $|x - x(t)| \ge 2r_0$. We consider the modified symbol in the following way.

$$n\lambda_{x_j}(t) - \lambda_{x_j}(x,t;D) = \left(n\lambda_{x_j}(t) - \lambda_{x_j}(\tilde{x}(x,t),t;D)\right)$$

$$+ \left(\lambda_{x_j}(\tilde{x}(x,t),t;D) - \lambda_{x_j}(x,t;D)\right).$$

For the first pseudo-differential operator on the right-hand side, we can apply sharp Gårding inequality. In fact,

$$\sup \left| n\lambda_{x_j}(t) - \lambda_{x_j}(\tilde{x}(x,t),t;\xi)\right| \le n\,Ar_0\,,$$

where sup is taken over $(x,\xi) \in \mathbb{R}^\ell \times \mathrm{supp}[\alpha_n(\xi)]$.

Next, for the second operator, observe first that the symbol $\lambda_{x_j}(\tilde{x}(x,t),t;\xi) - \lambda_{x_j}(x,t;\xi)$ vanishes when $x \in \mathrm{supp}[\beta_n]$. Thus, we commute $b(x,t;D)\alpha_n{}^{(p)}(D)$ with $\beta_{n_{(q)}}$ $(\, b = \lambda_{x_j}(\tilde{x},t;\xi) - \lambda_{x_j}(x,t;\xi)\,)$. Then, the commutator is estimated except for trivial factor by

$$\frac{1}{\nu!}\left|\left(b\alpha_n{}^{(p)}\right)^{(\nu)}\right|_0 \left|\beta_{n_{(q+\nu)}}\right|_0, \quad \text{for } |\nu| = N.$$

This is estimated, except for trivial factor, by

$$\left(\frac{N}{n}\, cr_0{}^{-1}\right)^{|p|}\left(Ncr_0{}^{-1}\right)^{|q|}\left(\frac{N^{1+2\varepsilon'}}{n}\,(cr_0{}^{-1})^2\right)^N.$$

By multiplying c^n_{pq}, and observing $1 + 2\varepsilon' < s$, we see easily (10). □

Proof of (12). We use the same notation as in the proof of Theorem A.1 in Lecture V.

$$C_{n,p,q}(u_n) = \sum_{2\le|\mu+\nu|\le N-(p+q)} \mu!^{-1}(-1)^{|\nu|}\nu!^{-1}\, a^{(\nu)}_{(\mu)}\left(\alpha_n^{(p+\mu)}\beta_{n\,(q+\nu)}\tilde{u}_n\right)$$

$$+ s_{N,p,q}\,\tilde{u}_n.$$

In view of the proof of Theorem A.1 in Lecture V, we see that

$$c^n_{pq}\|C_{n,p,q}(u_n)\| \le An \sum_{2\le|\mu+\nu|\le N-(p+q)} \left(2C_0|\mu|^{s-1}/n^{1-\frac{1}{s}}\right)^{|\mu|}$$

$$\left(2C_0|\nu|^{s-1}/n^{1-\frac{1}{s}}\right)^{|\nu|} \times c^n_{p+\mu,q+\nu}\|\alpha_n^{(p+\mu)}\beta_{n\,(q+\nu)}\tilde{u}_n\| + e^{-2/3\,N}.$$

First,

$$\left(2C_0|\mu|^{s-1}/n^{1-\frac{1}{s}}\right)^{|\mu|}\left(2C_0|\nu|^{s-1}/n^{1-\frac{1}{s}}\right)^{|\nu|} \le \left(2C_0|\mu+\nu|^{s-1}/n^{1-\frac{1}{s}}\right)^{|\mu+\nu|}.$$

Next, the number of (μ,ν) such that $|\mu+\nu| = m$ is estimated by $(m+1)^{2\ell-1}$. Third, the difference between $\sum c^n_{pq}\|\alpha_n^{(p)}\beta_{n\,(q)}u_n\|$ and $\sum c^n_{pq}\|\alpha_n^{(p)}\beta_{n\,(q)}\tilde{u}_n\|$ is negligible (see the proof of Theorem A.1).

Thus, we get

$$\sum_{|p+q|\le N} c^n_{pq}\|C_{n,p,q}(u_n)\| \le An\left(\sum_{2\le m\le N}\left(2C_0 m^{s-1}/n^{1-\frac{1}{s}}\right)^m (m+1)^{2\ell-1}\right)\overset{\circ}{S}_n$$

$$+ e^{-2/3\,N} \times \text{(some polynomial in } n) + \text{(negligible)}.$$

Now, if r_0 is small, we have $2C_0 m^{s-1}/n^{1-\frac{1}{s}} \le \frac{1}{2}$. Thus,

$$\sum \cdots \le \frac{\text{const.}}{n^{2(1-\frac{1}{s})}} \sum_{m=0}^{\infty}\left(\frac{1}{2}\right)^m (m+3)^{2\ell-1} \le \frac{\text{const.}}{n^{2(1-\frac{1}{s})}},$$

which proves (12).

§ 3. Some consequence of Theorem 1.

We explain how to apply Theorem 1 to the problems for differential operators. Let P be a differential operator, with coefficients in $\gamma^{(s)}$ $(s > 1)$, and with constant multiplicity. More precisely (see (6)~(8) in Lecture IV),

$$P = P_k \circ P_{k-1} \circ \cdots \circ P_1 + R \tag{16}$$

where

$$P_j = (\partial_t - i\lambda_j)^{m_j} + a_{1,j}(x,t;D)(\partial_t - i\lambda_j)^{m_j-1} + \cdots + a_{m_j,j}(x,t;D). \tag{17}$$

We assume

$$\rho_j \left(= \max_{1\le i\le m_j} \text{ order } a_{i,j}/_i \right) < \frac{1}{s}. \tag{18}$$

We make the same assumptions as in Lecture IV.

First we consider

$$P_j u = f. \tag{19}$$

Assume that $u(\cdot,t) \in C_t^m([0, \tau]; (\gamma^{(s)})')$. Let Γ be the bicharacteristic strip of P_j issuing from (x_0, ξ^0) at $t = 0$. More concretely, let $(x(t),\xi(t))$ be the solution of

$$\begin{cases} \dot{x}(t) = -\lambda_\xi(x,t;\xi) \\ \dot{\xi}(t) = \lambda_x(x,t;\xi) \end{cases}, \qquad (x(0),\xi(0)) = (x_0, \xi^0),$$

where $\lambda = \lambda_j$. Then

Theorem 2. Assume $(x(t),\xi(t)) \notin WF_s(f(\cdot,t))$ for $t \in [0, T]$. If

(A) $$(x_0, \xi^0) \underset{(\in)}{\notin} \bigcup_{i=0}^{m_j-1} WF_s\left(\partial_t^{\,i} u\Big|_{t=0}\right),$$

then it is the same with $u(\cdot,t)$ along Γ. Namely

(B) $$(x(t),\xi(t)) \underset{(\in)}{\notin} \bigcup_{i=0}^{m_j-1} WF_s\left(\partial_t^{\,i} u(\cdot,t)\right).$$

Proof. We reduce this theorem to Theorem 1. For simplicity we omit the suffix j, namely we write m, λ, a_i instead of m_j, λ_j, a_{ij}.

Put

$$(\partial_t - i\lambda)^j u = u_j \qquad (0 \le j \le m-1).$$

Then

$$\partial_t u_j = i\lambda u_j + u_{j+1} \qquad (0 \le j \le m-2)$$

$$\partial_t u_{m-1} = i\lambda u_{m-1} - \sum_{j=0}^{m-1} a_{m-j} u_j + f.$$

Next, denoting $<\Lambda> = \sqrt{1 + |D|^2}$, namely $\widehat{<\Lambda> u}(\xi) = (1 + |\xi|^2)^{1/2} \hat{u}(\xi)$, put

$$\tilde{u}_j = <\Lambda>^{(m-1-j)\rho} u_j \qquad (0 \le j \le m-1).$$

Then

$$\begin{cases} \partial_t \tilde{u}_j = i\lambda\tilde{u}_j + <\Lambda>^{\rho} \tilde{u}_{j+1} + [<\Lambda>^{(m-1-j)\rho}, i\lambda] <\Lambda>^{-(m-1-j)\rho} \tilde{u}_j \\ \partial_t \tilde{u}_{m-1} = i\lambda\tilde{u}_{m-1} - \sum a_{m-j} <\Lambda>^{-(m-1-j)\rho} \tilde{u}_j + f. \end{cases}$$

Recall that

$$\text{order}\left(a_{m-j}\langle\Lambda\rangle^{-(m-1-j)\rho}\right) \le \rho \qquad (0 \le j \le m-1).$$

(19) is equivalent to

$$\partial_t \tilde{U} = i\begin{pmatrix} \lambda & & & 0 \\ & \lambda & & \\ & & \ddots & \\ 0 & & & \lambda \end{pmatrix}\tilde{U} + A(x,t;D)\,\tilde{U} + F \tag{20}$$

where $\tilde{U} = {}^t(\tilde{u}_0, \cdots, \tilde{u}_{m-1})$, $F = {}^t(0, \cdots, 0, f)$, and

$$A = \left(a_{ij}(x,t;D)\right)_{1\le i,j\le m}, \quad \text{order } a_{ij} \le \rho. \tag{21}$$

In view of (21), we see that the same arguments as in the proof of Theorem 1 can be applied to (20). Observe the following fact:

$$(x(t),\xi(t)) \notin \bigcup_i WF_s\left(\partial_t^i u(\cdot,t)\right) \quad \text{and} \quad (x(t),\xi(t)) \notin \bigcup_i WF_s\left(\tilde{u}_i(\cdot,t)\right)$$

are equivalent in view of Lemma 5 in Lecture IV and Theorem A.1 in Lecture V. Thus the proof is complete. □

Now we return to

$$P\,u = f \tag{22}$$

where P, with coefficients in $\gamma^{(s)}_{x,t}$, satisfies (16) ~ (18). We can again apply the same arguments as in § 3 in Lecture IV. However, we should be careful for the reduction to a system. Namely, we use the perfect factorization in $\gamma^{(s)}$. We assume

$$\frac{1}{s} > \rho = \max(\rho_1, \cdots, \rho_k), \quad s > 1.$$

First, as in Lecture IV, we abtain a system of the following type:

$$\partial_t U = \begin{pmatrix} D_1 & 1 & \\ & D_2 & 1 \\ & & \ddots \end{pmatrix} U + R_1 U + F \tag{23}$$

where

$$D_j = \begin{pmatrix} i\lambda_j & 1 & & & 0 \\ & i\lambda_j & 1 & & \\ & & \ddots & \ddots & \\ & 0 & & \ddots & 1 \\ a_{m_j,j} & a_{m_j-1,j} & \cdots & & i\lambda_j + a_{1,j} \end{pmatrix}.$$

We can prove (this requires a little delicate arguments, however we omit its proof) that there exists a matrix M (order-1, with Gevrey pseudo-analytic symbol) such that (block diagonalization)

$$\partial_t (I + M) U = \begin{pmatrix} D_1 & & & \\ & D_2 & & \\ & & \ddots & \\ & & & D_k \end{pmatrix} (I + M) U + R_0 U + (I + M) F$$

where R_0 is a regularizing operator in $\gamma^{(s)}$, and $I + M$ is invertible (modulo regularizing operator) in $\gamma^{(s)}$:

$$(I + M)(I + M') = I + R_1, \quad (I + M')(I + M) = I + R_2$$

where R_i are also regularizing in $\gamma^{(s)}$.

We are now in a position to state Theorem 3. Let (x_0, ξ^0) and t_0 (>0) be given. Let $\Gamma_1, \cdots, \Gamma_k$ be the bicharacteristic strips issuing from (x_0, ξ^0) at $t = t_0$, corresponding to $\lambda_1, \cdots, \lambda_k$ respectively. Let (x_i, ξ^i) $(1 \le i \le k)$ be the values of Γ_i at $t = 0$.

<u>Theorem 3</u>. Let $u(\cdot,t)$ be the solution of (22) in $C_t^m\left((\gamma^{(s)})'\right)$, and suppose $f(\cdot,t) \in C_t^0(\gamma^{(s)})$. If for all j $(1 \le j \le k)$,

$$(x_j, \xi^j) \notin \bigcup_{i=0}^{m-1} WF_s\left(\partial_t^i u(\cdot,0)\right),$$

then

$$(x_0, \xi^0) \notin \bigcup_{i=0}^{m-1} WF_s\left(\partial_t^i u(\cdot,t_0)\right).$$

From this theorem, we obtain

<u>Corollary</u>. Let Σ be the characteristic conoid of the differential operator P with its vertex at $(x,t) = (y,0)$. Then the fundamental solution $E(t,x;y)$ for $t \ge 0$:

$$P\,E(t,x;y) = 0 \quad \text{for} \quad t > 0, \quad \partial_t^j E(0,x;y) = \begin{cases} 0 & (0 \le j \le m-2) \\ \delta_{x-y} & (j = m-1) \end{cases}$$

is C^∞ and belongs to $\gamma_{x,t}^{(s)}$ in the complement of Σ.

Proof. Let (x_0, t_0) $(t_0 > 0)$ be a point in the complementary set of Σ. Let further $\xi^0 \in \mathbb{R}^n \setminus \{0\}$ be an arbitrary point. Let $\Gamma_1, \cdots, \Gamma_k$ be the bicharacteristic strips corresponding to $\lambda_1, \ldots, \lambda_k$ respectively, issuing from (x_0, ξ^0) at $t = t_0$. Let $(x^1, \xi_1), \cdots, (x^k, \xi_k)$ be the value of $\Gamma_1, \cdots, \Gamma_k$ at $t = 0$. $(x_0, t_0) \notin \Sigma$ implies that $x^1, \cdots, x^k \neq y$ (otherwise, $(x_0, t_0) \in \Sigma$). Thus $\partial_t^i E(0,x;y) = 0$ $(0 \leq j \leq m-1)$ in each small neighborhood of $x^1, \cdots, x^k$. This implies, by Theorem,

$$(x_0, \xi^0) \notin \bigcup_{i=0}^{m-1} WF_s\left(\partial_t^i E(t_0, \cdot\,; y)\right).$$

Since ξ^0 is arbitrary, this proves $\partial_t^i E(t_0,x;y) \in \gamma_x^{(s)}$ $(0 \leq i \leq m-1)$ in a small neighborhood of x_0. Now we know that $E(t,x;y)$ is in $\gamma_{x,t}^{(s)}$ in a neighborhood of (x_0, t_0) (see § 4 in Lecture IV).

§ 4. Propagation of singularities in the sense of C^∞.

We explain our proof of Fundamental Theorem in Lecture V. As explained there, its proof is fairly simple. It consists of solving successively the transport equation. However we give here our proof, because this method works well even in the case where the differential operator P (with C^∞-coefficients) has constant multiplicity, and satisfies the Levi condition. Let us mention that this result is already known to Chazarain (see [4] in Lecture I). First we restate the Fundamental Theorem in Lecture V.

Let L be an elementary operator and consider the equation:

$$\begin{aligned} L(u) &\equiv \partial_t u + i\lambda(x,t;D)\,u + c(x,t;D)\,u \\ &= \partial_t u + a\,u = f\,, \end{aligned} \tag{24}$$

$$a = i\lambda + c$$

where $\lambda(x,t;\xi)$ is real and essentially homogeneous of degree 1 in ξ: $\lambda(x,t;\xi) \in S^1_{1,0}$, and $c(x,t;\xi) \in S^0_{1,0}$, depending smoothly on t.

<u>Theorem 4</u>. Assume $(x_0, \xi^0) \underset{(\in)}{\notin} WF(u(\cdot,0))$ and $(x(t),\xi(t)) \notin WF(f(\cdot,t))$ for $t \in [0, T]$. Then, for all $t \in [0, T)$,

$$(x(t),\xi(t)) \underset{(\in)}{\notin} WF(u(\cdot,t)).$$

In principle, our method is the same as in the case of Gevrey class. However, we use a slightly precise micro-localizer around Γ_n (= (x(t), $n\xi(t)$)). Instead of $(\alpha_n(\xi,t), \beta(x,t))$ defined by (4), we use $k_n(\xi,x,t)$ defined by

$$H_\lambda[k_n(\xi,x,t)] = 0.$$

Namely,

$$\begin{cases} \left(\partial_t + \lambda_\xi(x,t;\xi)\partial_x - \lambda_x(x,t;\xi)\partial_\xi\right) k_n(\xi,x,t) = 0 \\ k_n(\xi,x,0) = \alpha_n(\xi)\,\beta(x) \quad , \end{cases} \tag{25}$$

where $\{\alpha_n(\xi), \beta(x)\}$ is the same function in C^∞-case, defined in Lecture III and in § 4 in Lecture V, corresponding to (x_0, ξ^0). Let us observe that $k_n(\xi,x,t)$, in general, cannot be represented in the form $\alpha_n(\xi,t) \times \beta(x,t)$. Anyway, $k_n(\xi,x,t)$ has its support in a tubular domain around Γ_n.

To $k_n(\xi,x,t)$ (t being a parameter), we associate the pseudo-differential operator $k_n(D,x,t)$ defined by

$$\widehat{k_n(D,x,t)\,u}(\xi) = \int e^{-ix\xi}\, k_n(\xi,x,t)\, u(x)\, dx. \tag{26}$$

In other words,

$$k_n(D,x,t)\, u(x) = (2\pi)^{-\ell} \iint e^{i(x-y)\xi}\, k_n(\xi,y,t)\, u(y)\, dy\, d\xi. \tag{27}$$

The symbol of k_n is expressed by

$$\tilde{k}_n(x,t;\xi) = (2\pi)^{-\ell} \iint e^{-iy\eta}\, k_n(\xi+\eta,x+y,t)\, dy\, d\eta. \tag{28}$$

Observe the following fact. Let $a(\xi) \in C^{\infty}$, with at most polynomial growth when $|\xi| \to \infty$, and let $b(x) \in \mathcal{B}$. Then we have

$$a(D)\circ k_n(D,x,t)\circ b(x) = a(D)k_n(D,x,t)b(x)$$

<u>where $\circ$ stands for the operator product</u>. In other words, the symbol of the operator on the left-hand side is expressed by

$$(2\pi)^{-\ell}\iint e^{-iy\eta}\, a(\xi+\eta)\, k_n(\xi+\eta,x+y,t)\, b(x+y)\, dy\, d\eta\,.$$

Further, we have the following relation ($\longmapsto$ means the corresponding symbol),

$$a(x,D)\circ k_n(D,x,t) \longmapsto (2\pi)^{-\ell} \iint e^{-iy\eta}\, a(x,\xi+\eta)\, k_n(\xi+\eta,x+y,t)\, dy\, d\eta\,, \tag{29}$$

$$k_n(D,x,t)\circ a(x,D) \mapsto (2\pi)^{-\ell} \iint e^{-iy\eta}\, k_n(\xi+\eta, x+y, t)\, a(x+y, \xi)\, dy\, d\eta . \quad (30)$$

The proof will be given in Appendix.

Observe the following fact. In view of (25), if we take r_0 small, r_0 being the number which indicates the smallness of the supports of $\alpha_n(\xi)$ and $\beta(x)$, for $t \in [0, t_0]$ (t_0 being fixed), we have

i) $\operatorname{supp}[k_n(\xi,x,t)] \subset \{(\xi,x); |\xi - n\xi(t)| \leq nr_0', \ |x - x(t)| \leq r_0'\}$,

and $k_n(\xi,x,t) = 1$ for $|\xi - n\xi(t)| \leq \frac{1}{2} n r_0'$, $|x - x(t)| \leq r_0'/2$,

where r_0' becomes small with r_0. (31)

ii) $\left| k_{n\,(q)}^{(p)}(\xi,x,t) \right| \leq \dfrac{M_{pq}}{n^{|p|}}$.

$k_n(D,x,t)$ plays the same role as $\alpha_n(D,t)\beta_n(x,t)$ in § 1. In fact, using Theorem 3 ~ 4 in Lecture V, we can prove the following theorems.

In accordance with the expression of $S_{n,k}$ in § 4, Lecture V:

$$S_{n,k} = \sum_{|p+q|\leq k} n^{\frac{1}{2}|p|-\frac{1}{2}|q|} \| \alpha_n^{(p)} \beta_{(q)} u \| ,$$

we consider

$$S_{n,k} = \sum_{|p+q|\leq k} c_{pq}^{n} \| k_{n\,(q)}^{(p)} u \| , \qquad c_{pq}^{n} = n^{\frac{1}{2}|p|-\frac{1}{2}|q|} . \quad (32)$$

Theorem 5. Let $(x(t), \xi(t)) \notin WF(u(\cdot,t))$ for $t \in [0, t_0]$, where $t \mapsto u(\cdot,t) \in \mathcal{D}'_x$ is assumed to be continuous. Then, if r_0 is small (therefore r_0' is small in (31)), for arbitrary k,

$$S_{n,k}(t) = \sum_{|p+q| \le k} c_{pq}^{n} \| k_{n(q)}^{(p)}(D,x,t)\, u(\cdot,t) \|$$

is rapidly decreasing.

Theorem 6. t is fixed. If for arbitrary k,

$$\tilde{S}_{n,k}(t) = \sum_{|p| \le k} c_{p0}^{n} \| k_n^{(p)}(D,x,t)\, u(\cdot,t) \|$$

is rapidly decreasing, then $(x(t), \xi(t)) \notin WF(u(\cdot,t))$.

The proof will be given in Appendix.

Note. The term " rapidly decreasing ' is explained in § 4, Lecture V. We use also the term " negligible " in this case.

As usual, we use the decomposition:

$$1 = \chi_{n,-}(\xi,t) + \chi_n(\xi,t) + \chi_{n,+}(\xi,t) ,$$

its meaning will be clear. From (30), we see easily

Lemma 1.

1) $\| k_n{}^{(p)}_{(q)} u_{n,\pm} \| = \| k_n{}^{(p)}_{(q)} \chi_{n,\pm}(D,t)\, u(\cdot,t) \|$ is rapidly decreasing,

2) $\| [k_n{}^{(p)}_{(q)}, a]\, u_{n,\pm} \|$ is also rapidly decreasing.

Consider now

$$[k_n{}^{(p)}_{(q)}, a]\chi_n u = k_n{}^{(p)}_{(q)} \circ a\chi_n u - a \circ k_n{}^{(p)}_{(q)} \chi_n u .$$

Since $\chi_n(\xi,t) = 1$ for $\xi \in \operatorname{supp}_\xi[k_n(\xi,x,t)]$, denoting by $\tilde{\chi}_n(\xi,t)$ the characteristic function of $\operatorname{supp}_\xi[\chi_n(\xi,t)]$, and putting $\tilde{\chi}_n(D,t)\, u = \tilde{u}_n$, we see easily the following:

$$\| [k_n{}^{(p)}_{(q)}, a]\chi_n u - [k_n{}^{(p)}_{(q)}, a_n]\, \tilde{u}_n \|$$

is negligible, where $a_n = a(x,D)\chi_n(D)$.

Now we go back to the equation (24). Applying $k_n{}^{(p)}_{(q)}$ to (24), we get

$$(\partial_t + a)(k_n{}^{(p)}_{(q)} u) = g_{n,p,q} , \tag{33}$$

where

$$g_{n,p,q} = \partial_t k_n{}^{(p)}_{(q)} \tilde{u}_n - [k_n{}^{(p)}_{(q)}, a_n]\, \tilde{u}_n + k_n{}^{(p)}_{(q)} f + (\text{negligible}). \tag{34}$$

On the other hand, for (33), the following energy inequality holds.

$$\frac{d}{dt}\| k_{n\,(q)}^{\ (p)} u \| \leq \gamma \| k_{n\,(q)}^{\ (p)} u \| + \| g_{n,p,q} \| . \tag{35}$$

Hereafter we explain several elementary facts which we need for the estimate of $g_{n,p,q}$.

1) Asymptotic expression of $[k_{n\,(q)}^{\ (p)}, a_n(x,D)]$.

In view of (29) and (30), we obtain

$$[k_{n\,(q)}^{\ (p)}, a_n(x,D)] \sim \sum \mu!^{-1}(-1)^{|\nu|}\nu!^{-1}\, a_{n\,(\mu)}^{\ (\nu)} \circ k_{n\,(q+\nu)}^{\ (p+\mu)} . \tag{36}$$

For simplicity, we denote $k_{n\,(q)}^{\ (p)}\tilde{u}_n$ by $\tilde{u}_{pq}$. Then we have

$$\begin{aligned} c_{pq}^{\,n} \| [k_{n\,(q)}^{\ (p)}, a_n]\tilde{u}_n - \sum_{|\mu+\nu|=1} (-1)^{|\nu|} a_{n\,(\mu)}^{\ (\nu)} (\tilde{u}_{p+\mu,q+\nu}) \| \\ \leq C(N) \sum_{2\leq|\mu+\nu|\leq N} n^{1-\frac{1}{2}|\mu+\nu|}\, c_{p+\mu,q+\nu}^{\,n} \| \tilde{u}_{p+\mu,q+\nu} \| \\ + O\left(n^{-\frac{1}{2}|p+q|-N} \right) \| \tilde{u}_n \| . \end{aligned} \tag{37}$$

2) Another product $a(x,D) \odot b(D,x)$.

Let us recall the definition of $k_n(D,x,t)$ and the relation (25). In order to express this relation by operators, it is convenient to introduce the following product: To $a(x,\xi)$ and $b(\xi,x)$ we associate the pseudo-differential operator $a(x,D) \odot b(D,x)$, whose symbol is defined by

$$(2\pi)^{-\ell} \iint e^{-iy\eta}\, a(x+y, \xi+\eta)\, b(\xi+\eta, x+y)\, dy d\eta. \tag{38}$$

Then (25) implies

$$\partial_t k_n(D,x,t) + i \sum_j \lambda^{(e_j)} \odot k_{n(e_j)} - i \sum_j \lambda_{(e_j)} \odot k_n^{(e_j)} = 0.$$

3) Asymptotic expression of $a(x,D) \odot k_{n(q)}^{(p)}$.

In view of (38), it is easy to see that (assuming $a(x,\xi) \in S^m_{1,0}$),

$$a(x,D) \odot k_{n(q)}^{(p)} \sim a(x,D) \circ k_{n(q)}^{(p)} + \sum_{|\nu'+\nu''| \geq 1} \nu'!^{-1} \nu''!^{-1} a^{(\nu')}_{(\nu)} \circ k_{n(q)}^{(p+\nu'')} \tag{39}$$

$(\nu' + \nu'' = \nu)$.

There is some complexity in the consideration of $\partial_t k_{n(q)}^{(p)}$ $(|p+q| \geq 1)$. In fact, (25) yields by Leibniz,

$$\partial_t k_{n(q)}^{(p)} + i \sum_j \lambda^{(e_j)} k_{n(q+e_j)}^{(p)} - i \sum_j \lambda_{(e_j)} k_{n(q)}^{(p+e_j)}$$

$$= -i \sum_j \sum_{|p'+q'| \geq 1} \binom{p}{p'} \binom{q}{q'} \lambda^{(p'+e_j)}_{(q')} k_{n(q''+e_j)}^{(p'')} \tag{40}$$

$$+ i \sum_j \sum_{|p'+q'| \geq 1} \binom{p}{p'} \binom{q}{q'} \lambda^{(p')}_{(q'+e_j)} k_{n(q'')}^{(p''+e_j)},$$

where $p' + p'' = p$ and $q' + q'' = q$.

Now we put

$$\partial_t k_{n(q)}^{(p)} \tilde{u}_n + i \sum_j \lambda^{(e_j)} \circ k_{n(q+e_j)}^{(p)} \tilde{u}_n - i \sum_j \lambda_{(e_j)} \circ k_{n(q)}^{(p+e_j)} \tilde{u}_n = X_{n,p,q} \tilde{u}_n, \quad (41)$$

$$[k_{n(q)}^{(p)}, a_n] \tilde{u}_n + i \sum_j \lambda^{(e_j)} \circ k_{n(q+e_j)}^{(p)} \tilde{u}_n - i \sum_j \lambda_{(e_j)} \circ k_{n(q)}^{(p+e_j)} \tilde{u}_n = X'_{n,p,q} \tilde{u}_n. \quad (42)$$

Then we have (its proof is given in Appendix.)

<u>Lemma 2</u>.

$$\sum_{|p+q| \le k} c_{pq}^n \| X_{n,p,q} \tilde{u}_n \| , \qquad \sum_{|p+q| \le k} c_{pq}^n \| X'_{n,p,q} \tilde{u}_n \|$$

$$\le c(k) \sum_{|p+q| \le k} c_{pq}^n \| k_{n(q)}^{(p)} \tilde{u}_n \| + O(n^{-k/2}) \| \tilde{u}_n \| .$$

Since

$$g_{n,p,q} = X_{n,p,q} \tilde{u}_n - X'_{n,p,q} \tilde{u}_n + \text{(negligible)} + k_{n(q)}^{(p)} f ,$$

by summing up the energy inequality (35) with weight c_{pq}^n,

$$\frac{d}{dt} S_{n,k}(t) \le \gamma S_{n,k}(t) + \tilde{c}(k) \mathring{S}_{n,k}(t) + O(n^{-k/2}) \| \tilde{u}_n \| + c_{pq}^n \| k_{n(q)}^{(p)} f \|$$

where $\overset{\circ}{S}_{n,k}(t) = \sum_{|p+q|\le k} c^n_{pq} \| k_n{}^{(p)}_{(q)} \tilde{u}_n \|$. By Theorem 5, the last term is negligible, so $S_{n,k} - \overset{\circ}{S}_{n,k}$ is also negligible (essentially due to Lemma 1). Next, in view of the continuity $t \mapsto u(\cdot,t) \in \mathcal{D}'_x$, there exists an L (≥ 0) such that

$$\| \tilde{u}_n(\cdot,t) \| \le O(n^L), \quad \text{for} \quad t \in [0, T].$$

Let m be given, then we assign

$$k = 2m + 2L.$$

Thus we have

$$\frac{d}{dt} S_{n,k}(t) \le c'(k) S_{n,k}(t) + O(n^{-m}),$$

which implies

$$S_{n,k}(t) \le \exp(c't) \left(S_{n,k}(0) + t\, O(n^{-m}) \right) \le O(n^{-m}),$$

since $S_{n,k}(0)$ is negligible ($(x_0, \xi^0) \notin WF(u(\cdot,0))$). Now, m is arbitrary, which proves Theorem 4 by virtue of Theorem 6. □

Note As we mentioned at the beginning of this section, the same argument can be applied to the operator P with multiple characteristic satisfying Levi condition. This part is carried out along the same line as in § 3.

Appendix

§ A. 1. First we give the proof of (29) and (30). These are well known.

Proof of (29) We write $k(\xi,x)$ instead of $k_n(\xi,x,t)$.

$$a\,v(x) = (2\pi)^{-\ell}\int e^{ix\xi}\,a(x,\xi')\,\hat{v}(\xi)\,d\xi\,,$$

$$\hat{v}(\xi') = \int e^{-ix'\xi'}\,k(\xi',x')\,u(x')\,dx',$$

$$u(x') = (2\pi)^{-\ell}\int e^{ix'\xi}\,\hat{u}(\xi)\,d\xi\,.$$

Thus,

$$(a\circ k)\,u(x) = (2\pi)^{-\ell}\iiint e^{ix\xi'-ix'\xi'+ix'\xi}\,a(x,\xi')k(\xi',x')\hat{u}(\xi)\;dx'd\xi'd\xi.$$

Hence, the symbol of $a\circ k$ is expressed by

$$(2\pi)^{-\ell}\iint e^{-ix\xi+ix\xi'-ix'\xi'+ix'\xi}\,a(x,\xi')k(\xi',x')\;dx'd\xi'$$

$$= (2\pi)^{-\ell}\iint e^{-i(x'-x)(\xi'-\xi)}\,a(x,\xi')k(\xi',x')\;dx'd\xi'$$

$$= (2\pi)^{-\ell}\iint e^{-iy\eta}\,a(x,\xi+\eta)\,k(\xi+\eta,x+y)\;dyd\eta.$$

Proof of (30).

$$k\,v(x) = (2\pi)^{-\ell} \iint e^{i(x-x')\xi'}\, k(\xi',x')\, v(x')\, dx'd\xi',$$

$$a\,u(x') = (2\pi)^{-\ell} \int e^{ix'\xi}\, a(x',\xi)\, \hat{u}(\xi)\, d\xi.$$

Hence

$$(k\circ v)\, u(x) = (2\pi)^{-2\ell} \iiint e^{i(x-x')\xi'+ix'\xi}\, k(\xi',x')a(x',\xi)\hat{u}(\xi)\, dx'd\xi'd\xi.$$

We see that the symbol of $k\circ a$ is expressed by

$$(2\pi)^{-\ell} \iint e^{-ix\xi+i(x-x')\xi'+ix'\xi}\, k(\xi',x')a(x',\xi)\, dx'd\xi'$$

$$= (2\pi)^{-\ell} \iint e^{-i(x'-x)(\xi'-\xi)}\, k(\xi',x')a(x',\xi)\, dx'd\xi$$

$$= (2\pi)^{-\ell} \iint e^{-iy\eta}\, a(\xi+\eta,x+y)\, a(x+y,\xi)\, dyd\eta. \qquad \square$$

Proof of Theorem 5.

We define $\alpha_n(\xi,t)$, $\beta(x,t)$ in the following way: Let $\alpha(\xi) \in C_0^\infty$, $\mathrm{supp}[\alpha] \subset \{\xi;\ |\xi| \le \tilde{r}_0\}$, $\alpha(\xi) = 1$ for $|\xi| \le \frac{1}{2}\tilde{r}_0$. Putting $\alpha_n(\xi) = \alpha(\frac{\xi}{n})$ we define $\alpha_n(\xi,t) = \alpha_n(\xi - n\,\xi(t))$. In the same way, $\beta(x) \in C_0^\infty$, $\mathrm{supp}[\beta] \subset \{x;\ |x| \le \tilde{r}_0\}$, $\beta(x) = 1$ for $|x| \le \frac{1}{2}r_0$. We define $\beta(x,t) = \beta(x - x(t))$.

By Theorem 3 in Lecture V, the assumption implies the following: If $\tilde{r}_0$ is chosen small, then for any k,

$$\tilde{S}_{n,k}(t) = \sum_{|p+q|\le k} c^{n}_{pq} \| \alpha_n^{(p)}(D,t)\beta_{(q)}(x,t)\, u(\cdot,t) \|$$

is uniformly and rapidly decreasing with respect to $t \in [0, t_0]$.

Now if $r_0' \le \frac{1}{2}\tilde{r}_0$, then $k_n(\xi,x,t) \Subset \alpha_n(\xi,t)\ \beta(x,t)$, $\forall t \in [0, t_0]$.
In the consideration of $\| k_{n\,(q)}^{(p)} u(\cdot,t)\|$, it is enough to consider u_n instead of u (Lemma 1).

$$k_{n\,(q)}^{(p)} u_n = \alpha_n \circ k_{n\,(q)}^{(p)} \circ \beta u_n = \sum_{|\mu|\le N} \mu!^{-1} k_{n\,(q+\mu)}^{(p)} \circ \alpha_n^{(\mu)}(\beta u_n) + r_{N,p,q}(\beta u_n).$$

First,

$$\| r_{N,p,q}\| \le O(n^{-|p|-N-1}) \quad \text{and} \quad \| \beta u_n \| \le O(n^{L}) \quad (\exists L > 0).$$

Hence

$$\| k_{n\,(q)}^{(p)} u_n \| \le c(N) \sum_{|\mu|\le N} n^{-|p|} \| \alpha_n^{(\mu)} \beta u_n \| + O(n^{-|p|-N-1+L}).$$

Since the difference $\| \alpha_n^{(\mu)} \beta u_n \| - \| \alpha_n^{(\mu)} \beta u \|$ is negligible, and that N can be taken large as we wish, we see that $\| k_{n\,(q)}^{(p)} u_n \|$, hence $\| k_{n\,(q)}^{(p)} u \|$ is rapidly decreasing. □

Proof of Theorem 6.

We reduce this fact to Theorem 4 in Lecture V. We take $\alpha_n(\xi)$, $\beta(x)$ as usual, such that

$$\alpha_n(\xi)\,\beta(x) \Subset k_n(\xi,x,t)\,.$$

Namely, we suppose that on $\mathrm{supp}[\alpha_n(\xi)\,\beta(x)]$, $k_n(\xi,x,t) = 1$. We use Lemma 1, 1), so that it is enough to consider $\tilde{u}_n$ ($= \tilde{\chi}_n u$) (see the beginning part after Lemma 1) instead of u itself.

$$\alpha_n^{(p)}\beta\tilde{u}_n = \alpha_n^{(p)}k_n(D,x,t)\beta\tilde{u}_n = \sum_{|\nu|\le N} (-1)^{|\nu|}\nu!^{-1}\alpha_n^{(p)}\beta_{(\nu)}k_n^{(\nu)}\tilde{u}_n + r_{N,p}\tilde{u}_n.$$

Thus,

$$\|\alpha_n^{(p)}\beta\tilde{u}_n\| \le c(N) \sum_{|\nu|\le N} n^{-p}\|k_n^{(\nu)}\tilde{u}_n\| + O(n^{-p-N-1})\|\tilde{u}_n\|\,.$$

Since $\|\tilde{u}_n\| \le O(n^L)$ ($\exists\, L \ge 0$), and that we can take N arbitrarily large, we see that $\|\alpha_n^{(p)}\beta u_n\|$, hence $\|\alpha_n^{(p)}\beta u\|$ is rapidly decreasing. □

§ A. 2. Proof of Lemma 2.

The second inequality of Lemma 2 is almost evident from (37). In fact, since $a = i\lambda + c$,

$$\sum_{|\mu+\nu|=1} (-1)^{|\nu|} a_{(\mu)}^{(\nu)}\tilde{u}_{p+\mu,q+\nu}$$

$$= i\sum_j \lambda_{(e_j)}\circ k_{n\,(q)}^{(p+e_j)}\tilde{u}_n - i\sum_j \lambda^{(e_j)}\circ k_{n\,(q+e_j)}^{(p)}\tilde{u}_n$$

$$+ \sum_{|\mu+\nu|=1} (-1)^{|\nu|} c_{(\mu)}^{(\nu)}(\tilde{u}_{p+\mu,q+\nu}),$$

and

$$c^n_{pq} \sum_{|\mu+\nu|=1} \| c^{(\nu)}_{(\mu)} \tilde{u}_{p+\mu,q+\nu} \| \le \text{const.} \sum_{|\mu+\nu|=1} n^{-\frac{1}{2}} c^n_{p+\mu,q+\nu} \| \tilde{u}_{p+\mu,q+\nu} \| .$$

If we take in (37) $N = k - |p + q|$, the last term of (37) becomes

$$O(n^{\frac{1}{2}|p+q|-k}) \| \tilde{u}_n \| \le O(n^{-k/2}) \| \tilde{u}_n \| \qquad (\text{since } |p + q| \le k).$$

The first inequality of Lemma 2 is fairly delicate. In view of (39) and (40), it is enough to consider the following terms:

$$\lambda^{(e_j)} \odot k_n{}^{(p)}_{(q+e_j)} - \lambda^{(e_j)} \circ k_n{}^{(p)}_{(q+e_j)} \equiv A,$$

$$\lambda_{(e_j)} \odot k_n{}^{(p+e_j)}_{(q)} - \lambda_{(e_j)} \circ k_n{}^{(p+e_j)}_{(q)} \equiv B,$$

$$\lambda^{(p+e_j)}_{(q')} \odot k_n{}^{(p'')}_{(q''+e_j)} \equiv C,$$

$$\lambda^{(p')}_{(q'+e_j)} \odot k_n{}^{(p''+e_j)}_{(q'')} \equiv D, \qquad |p' + q'| \ge 1.$$

First consider $A \tilde{u}_n$. By virtue of (39),

$$c^n_{pq} \| A \tilde{u}_n \| \le \text{const.} \sum_{1 \le |\nu'+\nu''| \le N'} c^n_{pq} \left\| \lambda^{(e_j+\nu')}_{(\nu)} \circ k_n{}^{(p+\nu'')}_{(q+e_j)} \tilde{u}_n \right\|$$

$$+ O(n^{-\frac{1}{2}|p+q|-N'-1}) \| \tilde{u}_n \| .$$

The first term is estimated by

$$\sum_{1\le|\nu|\le N'} n^{-|\nu'|-\frac{1}{2}|\nu''|+\frac{1}{2}} c^n_{p+\nu'',q+e_j} \| \tilde{u}_{p+\nu'',q+e_j} \| .$$

We choose $N' = k - |p + q| - 1$. Then

$$c^n_{pq} \| A \tilde{u}_n \| \le c(N') \sum_{|p+q|\le k} c^n_{pq} \| \tilde{u}_{p,q} \| + O(n^{-k/2}) \| \tilde{u}_n \| .$$

For $c^n_{pq} \| B \tilde{u}_n \|$, we get the same estimate.

Now we consider the third term. By using (39),

$$c^n_{pq} \left\| \lambda^{(p'+e_j)}_{(q')} \odot k_n{}^{(p'')}_{(q''+e_j)} \tilde{u}_n \right\| \le c^n_{pq} \text{ const.} \sum_{|\nu|\le N'} \left\| \lambda^{(p'+e_j+\nu')}_{(q'+\nu)} \circ k_n{}^{(p''+\nu'')}_{(q''+e_j)} \tilde{u}_n \right\|$$

$$+ O(n^{-\frac{1}{2}|p+q|-N'}) \| \tilde{u}_n \| .$$

Now, general term in the first part is estimated by

$$\text{const. } n^{-|p'|-|\nu'|} \frac{c^n_{pq}}{c^n_{p''+\nu'',q''+e_j}} \cdot c^n_{p+\nu'',q''+e_j} \| \tilde{u}_{p''+\nu'',q''+e_j} \|$$

$$\le \text{const. } n^{-\frac{1}{2}|p'+q'|-|\nu'|-\frac{1}{2}|\nu''|+\frac{1}{2}} c^n_{p+\nu'',q''+e_j} \| \tilde{u}_{p''+\nu'',q''+e_j} \| .$$

We take $N' = k - |p + q|$. Taking account of $|p' + q'| \ge 1$, we get

$$\| C u_n \| \le c(N') \sum_{|p+q|\le k} c^n_{pq} \| \tilde{u}_{pq} \| + O(n^{-k/2}) \| \tilde{u}_n \| .$$

It is the same with $D \tilde{u}_n$.

Lecture VII Schrödinger Type Equations

§ 1. Introduction (General view-points on evolution equations).

The hyperbolic equation has been investigated in great detail from th 70 th. This equation is very important in view of its applications, moreover many important mathematical concepts are related to the problems on hyperbolic equations. However we should recall here three types equations which we touched upon in the first lecture. Let us explain them from a more general view-point.

Let the evolution equation be

$$\begin{cases} \partial_t^m u(x,t) + \sum_{j=1}^{m} a_j(x,t;\partial x)\, \partial_t^{m-j} u(x,t) = f(x,t) \\ \partial_t^j u\Big|_{t=0} = u_j(x) \qquad (0 \le j \le m-1). \end{cases} \tag{1}$$

The hyperbolic equation could be characterized by the properties

i) It has the property of finite propagation speed;

ii) It is C^∞-wellposed not only for positive direction in t, but also for negative direction.

The property i) says the following. Suppose always $f = 0$. There exists a strict forward cone C with vertex at the origin, say

$$C = \{(x,t);\ |x| \le \lambda_{max} t,\ \ t \ge 0\},$$

such that the equation $u(x,t)$ of (1) for $t \ge t_0$ (≥ 0) has its support within

$$\bigcup_{j=0}^{m-1} \operatorname{supp}[\partial_t^j u(\cdot,t_0)] + C .$$

We know when (1) has this property, then

$$\text{order } a_j \leq j$$

must hold (Mizohata [7]). Such an equation is called Kovalevskian, so that every evolution equation with finite propagation speed is Kovalevskian. We consider non-Kovalevskian evolution equation.

i) parabolic type equation

This could be characterized by the following properties:

a) It is H^{∞}- wellposed for positive direction in t, but never H^{∞}-wellposed for negative direction.

b) As an operator it is hypoelliptic.

ii) Schrödinger type equation

This equation is H^{∞}-wellposed for both directions in t.

For non-Kovalevskian equations, the investigations (characterizations) are delicate problems. There are few results. The author is feeling that now it is the time to attack this problem.

We discuss here a simple Schrödinger type equation. The Schrödinger equation has the form:

$$i\,\partial_t u + \Delta u + c(x)\,u = f.$$

Many papers have been published already in the case $n = 3$. However, in the case of N particles with strong interactions, namely in the case $n = 3N$, the treatment is quite different from the case $n = 3$, and it becomes considerably difficult, and essential progress has just started.

Here we discuss the L^2-wellposedness of

$$L(u) \equiv i\,\partial_t u + \Delta u + \sum_{j=1}^{n} b_j(x)\partial_{x_j} u + c(x)\,u = f. \tag{2}$$

We say that the Cauchy problem for (2) is L^2-wellposed, if for any initial data $u_0(x) \in L^2(\mathbb{R}^n)$, and for $f(\cdot,t) \in C_t^0(L^2(\mathbb{R}^n))$, there exists a unique solution $u(x,t)$ for $|t| \le T$. Moreover we assume that the following type of estimate holds:

$$\|u(\cdot,t)\| \le C(T)\left(\|u(\cdot,0)\|_{L^2} + \left|\int_0^t \|f(\cdot,s)\|_{L^2}\,ds\right|\right), \quad |t| \le T. \tag{3}$$

In the case $n = 1$, as mentioned in Lecture II, in 1974, Takeuchi presented a sufficient condition [8]. Let us explain this. In this case (2) becomes

$$i\,\partial_t u + \partial_x^2 u + b(x)\partial_x u + c(x)\,u = f.$$

Let us assume all the coefficients are bounded with all derivatives. Take $p(x)\,u$ instead of u. If we choose $p(x)$ suitably, we can eliminate first order term by the following consideration. Since

$$p(x)\partial_x^2 u = \partial_x^2(p(x)\,u) - 2\,p'(x)\,\partial_x u - p''(x)\,u,$$

we choose $p(x)$ in such a way that $p(x)\,b(x) - 2\,p'(x) = 0$. Namely,

$$p(x) = \exp\left(\frac{1}{2}\int_0^x b(y)\,dy\right).$$

Then the equation becomes

$$i\,\partial_t(pu) + \partial_x^2(pu) + \tilde{c}(x)(pu) = p(x)\,f$$

where $\tilde{c}(x) = c(x) - p''(x)/p(x)$. Thus, if $p(x)$ satisfies

$$0 < {}^{\exists}c \le |p(x)| \le {}^{\exists}M,$$

then the above equation is L^2-wellposed. Now this condition is equivalent to

$$\left| \mathrm{Re} \int_0^x b(y)\, dy \right| \le {}^{\exists}M' \quad \text{for all} \quad x \in \mathbb{R}. \tag{4}$$

Observe that in the case $b(x)$ is constant, we see easily that $\mathrm{Re}\, b = 0$ is necessary and sufficient. The condition (4) says that $\mathrm{Re}\, b(x) \equiv 0$ is not necessary, whereas this is concerned with a property of a <u>global</u> behavior of $b(x)$, which the micro-local analysis (at least up to now) dared not to consider. In 1980, Takeuchi investigated (2), and presented the following condition ([9]), which is a natural extension of (4):

$$(C_0) \begin{cases} {}^{\forall}\omega \in S^{n-1}, \quad {}^{\forall}(x,t) \in \mathbb{R}^n \times \mathbb{R}, \\ \\ \mathrm{Re} \displaystyle\int_0^t \sum_{j=1}^{n} b_j(x + \omega \cdot s)\, \omega_j\, ds \quad \text{remains bounded.} \end{cases}$$

Takeuchi claimed that this condition is a sufficient condition for the L^2-wellposedness. However, it seems to the author his proof is not clear. In spite of this, the condition (C_0) which Takeuchi proposed has its own beauty and attracts our interest. Even now, we do not know whether his claim is correct or not. Here we show first that the condition (C_0) is a necessary condition. After then, we present a sufficient condition. The result which we show here is the same as in [6].

For this purpose, we borrow the idea from the paper of G.D.Birkhoff [1], which discussed the relation between the Schrödinger equation and the Newton equation. In this paper, he showed that the Schrödinger equation can be obtained as a first approximation when the Planck constant $\hbar$ becomes small. Let us observe that this idea was extensively developed by Maslov [5] and Leray [4].

§ 2. Necessity of (C_0).

Put

$$u(x,t) = e^{i\psi(x,t;\xi)} v(x,t;\xi) .$$

Then

$$e^{-i\psi} L[e^{i\psi} v] = - [\partial_t \psi + \sum_{j=1}^{n} (\partial_{x_j} \psi)^2] v + i T_\psi [v] + M[v],$$

where

$$\begin{cases} T_\psi = \partial_t + 2 \sum_{j=1}^{n} \psi_{x_j} \cdot \partial_{x_j} + \sum_j \psi_{x_j} b_j + \Delta \psi \\ M = \Delta + \sum_j b_j \partial_{x_j} + c. \end{cases}$$

T_ψ is called transport equation. First we define ψ (phase function) as a solution of

$$\partial_t \psi + \sum_j (\partial_{x_j} \psi)^2 = 0.$$

Here we take

$$\psi(x,t;\xi) = x \cdot \xi - |\xi|^2 t .$$

Observe that $\Delta \psi = 0$. Thus the above equation becomes

$$e^{-i\psi} L[e^{i\psi} v] = i T_\psi [v] + M[v]. \tag{5}$$

We define v as a solution of $T_\psi [v] = 0$. Namely,

$$\left(\partial_t + 2 \sum_j \xi_j \partial_{x_j} + \sum_j b_j(x) \xi_j \right) v = 0. \tag{6}$$

If we assign $v\Big|_{t=0} = \Psi(x)$, then

$$v(x,t;\xi) = \exp\left[-\int_0^t \sum_j b_j(x - 2\xi s)\xi_j\, ds\right]\Psi(x - 2\xi t). \tag{7}$$

We want to show that <u>if (C_0) is violated, then the inequality (3) fails to hold</u>. For this purpose, observe the following relations.

i) $$\int_0^{-t} \sum_j b_j(x + \omega s)\omega_j\, ds = \int_0^t \sum_j b_j(x - \omega s)(-\omega_j)\, ds,$$

ii) $$\int_0^t \sum_j b_j(x + \omega s)\omega_j\, ds = -\int_0^t \sum_j b_j(x' - \omega s)(-\omega_j)\, ds,$$

where $x' = x + \omega t$,

iii) for $\rho > 0$,

$$\int_0^t \sum_j b_j(x - \omega s)\omega_j\, ds = \int_0^{\frac{t}{\rho}} \sum_j b_j(x - \rho\omega\cdot s)\rho\omega_j\, ds.$$

By these properties, if (C_0) fails to hold, there exist t_0 (> 0), $\omega^0 \in S^{n-1}$, and $x^1 \in \mathbb{R}^n$, such that

$$-\operatorname{Re}\int_0^{t_0} \sum_j b_j(x^1 - 2\omega^0 s)\omega_j^0\, ds \geq \log 2C(T), \tag{8}$$

where $C(T)$ is the same constant as in (3).

Put

$$\begin{cases} x^1 - 2\omega^0 t_0 = x^0 \\ t_0/_\rho = t_\rho, \end{cases}$$

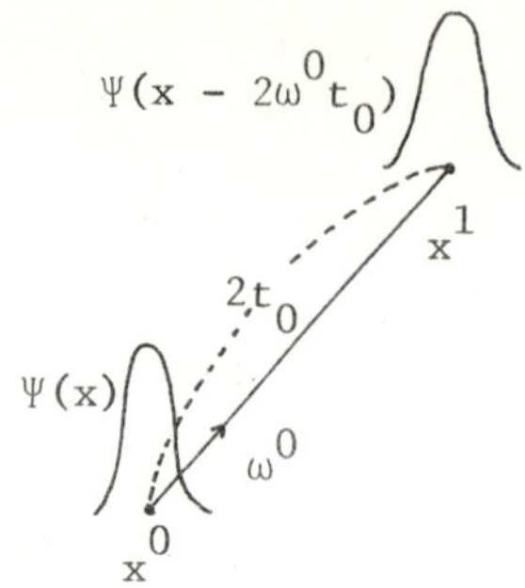

Fig.

In view of iii), (7) can be written as

$$v(x^1, t_\rho; \rho\omega^0) = \exp\left[-\int_0^{t_0} \sum_j b_j(x^1 - 2\omega^0 s)\omega_j{}^0 ds\right] \psi(x^0),$$

$$v(x, t_\rho; \rho\omega^0) = \exp\left[-\int_0^{t_0} \sum_j b_j(x - 2\omega^0 s)\omega_j{}^0 ds\right] \psi(x - 2\omega^0 t_0).$$

We take $\psi(x)$ <u>with small support around $x = x^0$</u>, satisfying $\int |\psi|^2 dx = 1$. (See Fig.) Thus, in view of (8),

$$\left\| v(\cdot, t_\rho; \rho\omega^0) \right\|_{L^2(\mathbb{R}^n)} \geq \frac{3}{2} C(T). \tag{9}$$

(Observe that $\psi(x - 2\omega^0 t_0)$ has small support around $x = x^1$.) Moreover, $M[v(x, t; \rho\omega^0)]$ is uniformly bounded for $t \in [0, t_\rho]$ when $\rho \to \infty$. This implies

$$\int_0^{t_\rho} \| M[v(\cdot, s; \rho\omega^0)] \| ds \to 0 \quad (\rho \to \infty). \tag{10}$$

On the other hand, in view of (5), $L[e^{i\psi} v] = e^{i\psi} M[v]$, and $e^{i\psi} v\Big|_{t=0} = e^{ix\rho\omega^0} \psi(x)$, thus its L^2-norm is 1. These, together with (9) and (10), show that the inequality (3) fails to holds.

§ 3. Sufficiency for L^2-wellposedness.

We try to solve the equation (2) under the assumption (C_0). The basic idea is to eliminate the first order term. For this purpose, put

$$u(x,t) = k(x,t;D)\, v(x,t)\,, \quad \text{with} \quad k(x,0;\xi) = 1.$$

Then

$$L(u) = (2\pi)^{-n} \int e^{ix\xi}[i\partial_t + \sum_j (i\xi_j + \partial_{x_j})^2 + \sum_j b_j(i\xi_j + \partial_{x_j}) + c] \times$$
$$\times \left(k(x,t;\xi)\, \hat{v}(\xi,t)\right) d\xi.$$

Define $k(x,t;\xi)$ by

$$\left(\partial_t + 2\sum_j \xi_j\partial_{x_j} + \sum_j b_j(x)\xi_j\right)k(x,t;\xi) = 0,$$
$$k(x,0;\xi) = 1. \tag{11}$$

Then $L(u) = f$ becomes

$$k\,(i\partial_t + \Delta)\, v + k_1(x,t;D)\, v = f, \tag{12}$$

where

$$k_1(x,t;\xi) = M[\,k\,] = \left(\Delta + \sum_j b_j(x)\partial_{x_j} + c\right) k(x,t;\xi).$$

From (11), we get

$$\begin{cases} k(x,t;\xi) = \exp[\,\psi(x,t;\xi)\,], \\ \psi(x,t;\xi) = -\displaystyle\int_0^t \sum_j b_j(x - 2\xi s)\,\xi_j\, ds. \end{cases}$$

The next step is to show that, under certain additional conditions,

i) $k(x,t;D)$, $k_1(x,t;D)$ are bounded operators in $L^2(\mathbb{R}^n)$,

ii) If $|t|$ is small, say $|t| \leq t_0$, $k(x,t;D)$ is invertible in $L^2(\mathbb{R}^n)$.

If this is realized, then (12) becomes

$$(i\partial_t + \Delta + k^{-1} \circ k_1)\, v = k^{-1} f.$$

This shows that (2) is solvable for $|t| \leq t_0$, and as we shall show it, if t_0 is independent of initial data, we can repeat the argument and we can conclude that (2) is solvable for any t.

Now we present a sufficient condition to guarantee the above properties i) and ii).

$$(C_\alpha) \begin{cases} \forall \alpha, \ |\alpha| \geq 1, \\ \displaystyle\int_0^\infty |\partial_x^\alpha b_j(x + \omega s)|\, ds \quad (j = 1,2,\cdots,n) \ \text{remains bounded when} \\ (x,\omega) \ \text{runs through} \ \mathbb{R}^n \times S^{n-1}. \end{cases}$$

Assume (C_0) and (C_α). Then

$$\left| \partial_\xi^\alpha \partial_x^\beta k(x,t;\xi) \right| \leq M_{\alpha\beta} |t|^{|\alpha|}. \tag{13}$$

This implies that k and k_1 are bounded operators by Calderón-Vaillancourt Theorem. Next define the operator $e^{-\psi(x,t;D)}$ by its symbol $e^{-\psi(x,t;\xi)}$. We see easily that this is also a bounded operator by (C_α). Moreover, if we put

$$k(x,t;D)\circ e^{-\psi} = I + r(x,t;D),$$

$$e^{-\psi}\circ k(x,t;D) = I + r'(x,t;D),$$

we shall show that

$$\| r(x,t;D)\| , \| r'(x,t;D)\| < 1, \quad \text{for} \quad |t| \le t_0. \tag{14}$$

In order to see this, let $p(x,\xi)$ and $q(x,\xi)$ be two symbols; $p\circ q$ stands for its operator product, and $p\cdot q$ stands for its symbol product. Put

$$r(x,D) = (p\circ q)(x,D) - (p\cdot q)(x,D).$$

Then,

$$(p\circ q)(x,\xi) = (2\pi)^{-n} \iint e^{-iy\eta} p(x,\xi+\eta) q(x+y,\xi)\, dy\, d\eta .$$

This implies

$$r(x,\xi) = \int_0^1 S_\theta(x,\xi)\, d\theta,$$

$$S_\theta(x,\xi) = (2\pi)^{-n} \iint e^{-iy\eta}\, i^{-1} \sum_{j=1}^{n} p_{\xi_j}(x,\xi+\theta\eta) q_{x_j}(x+y,\xi)\, dy\, d\eta .$$

We apply this formula to the case $p = e^{\psi}$, $q = e^{-\psi}$ or $p = e^{-\psi}$, $q = e^{\psi}$ (recall $p\cdot q = 1$). Then, by virtue of (13), and by Calderón-Vaillancourt Theorem, we see easily (14).

Summing up:

Theorem. Suppose that (2) is L^2-wellposed. Then (C_0) is a necessary condition. If we assume moreover (C_α), then (2) becomes L^2-wellposed.

Note. Recently, Ichinose succeeded in extending the above result to the case of H^∞-wellposedness ([2], [3]). In this case,

$$(C_0)' \qquad \left| \operatorname{Re} \int_0^t \sum_{j=1}^n b_j(x + \omega\cdot s)\,\omega_j \, ds \right| \le C_1 \log(1 + |t|) + C_2$$

$(\exists\, C_1, C_2)$ becomes a necessary condition.

References

[1] G.D. Birkhoff, Quantum mechanics and asymptotic series, Bull. Amer. Math. Soc. 39(1939), 681-700.

[2] M.Ichinose, Sufficient condition on H^{∞}-wellposedness for Schrödinger type equations, Comm. in P.D.E., 9(1984), 33-48.

[3] ——————, Some remarks on the Cauchy problem for Schrödinger type equations, to appear in Osaka J. Math.

[4] J. Leray, Lagrangian analysis and quantum mechanics, 1981, M.I.T.

[5] V.P. Maslov, Théorie des perturbations et méthodes asymptotiques (French translation), 1972.

[6] S. Mizohata, Sur quelques équations du type Schrödinger, Séminaire de Vaillant à Paris VI (1981).

[7] ——————, On evolution equations with finite propagation speed, Israel J. Math. 13(1972), 173-187.

[8] J. Takeuchi, A necessary condition for the well-posedness of the Cauchy problem for a certain class of evolution equations, Proc. Japan Acad. 50(1974), 133-137.

[9] ——————, On the Cauchy problem for some non-kowalewskian equations with distinct characteristic roots, J. Math. Kyoto Univ., 20(1980), 105-129.